CIRCUIT THEORY
VOLUME TWO

ELECTRONIC AND ELECTRICAL ENGINEERING TEXTS

Edited by

Professor P. J. B. Clarricoats
Queen Mary College, University of London

and

Professor P. J. Lawrenson
University of Leeds

---

1. Systems of Units in Electricity and Magnetism   *LEO YOUNG*
2. Electrical Power Systems Vol. 1   *A. E. GUILE* and *W. PATERSON*
3. Digital Computer Design   *F. G. HEATH*
4. Electromagnetic Theory Vol. 1   *J. B. DAVIES* and *D. E. RADLEY*
5. Theory of Communication   *A. E. KARBOWIAK*
6. Circuit Theory Vol. 1   *J. O. SCANLAN* and *R. LEVY*
7. Control Theory   *C. B. SPEEDY, R. F. BROWN* and *G. C. GOODWIN*
8. Electrical Power Systems Vol. 2   *A. E. GUILE* and *W. PATERSON*
9. Electromagnetic Theory Vol. 2   *J. B. DAVIES* and *D. E. RADLEY*

J. O. SCANLAN
M.E., Ph.D., D.Sc., C.Eng., F.I.E.E.
Sen. Mem. I.E.E.E., F.I.M.A.

*Professor of Electronic Engineering,
University College, Dublin,
Ireland.
Formerly
Professor of Electronic Engineering,
University of Leeds.*

R. LEVY
M.A., Ph.D., C.Eng., M.I.E.E.
Sen. Mem. I.E.E.E.

*Vice-President for Research,
Microwave Development Laboratories, Inc.,
Natick, Massachusetts, U.S.A.*

CIRCUIT THEORY
VOLUME TWO

OLIVER & BOYD · EDINBURGH

OLIVER & BOYD
Croythorn House
23 Ravelston Terrace
Edinburgh EH4 3TJ
A division of Longman Group Ltd

ISBN 0 05 002687 9
First published 1973
© 1973 J. O. Scanlan and R. Levy
All rights reserved

No part of this publication may be reproduced, stored in a retrieval system, or transmitted in any form or by any means—electronic, mechanical, photocopying, recording or otherwise—without the prior permission of the Copyright Owners and the Publisher. The request should be addressed to the Publisher in the first instance.

Printed in Great Britain by
Bell & Bain Ltd., Glasgow

CONTENTS

PREFACE ix

Chapter 7 THE COMPLEX FREQUENCY PLANE
- 7.1. Introduction — 285
- 7.2. The Definition of Complex Natural Frequencies — 288
- 7.3. Poles and Zeros — 289
- 7.4. Real Frequency Behavior from Pole-zero Locations — 291
  - 7.4.1. Series RL Circuit — 291
  - 7.4.2. Series RLC Circuit — 292
- 7.5. General Circuit Analysis Using Complex Frequency — 296
  - 7.5.1. Mesh or Loop Analysis — 296
  - 7.5.2. Nodal Analysis — 303
- 7.6. Driving Point and Transfer Functions — 306
  - 7.6.1. Properties of Driving Point and Transfer Functions — 309
  - 7.6.2. Special Properties of Driving Point Functions — 311
  - 7.6.3. Special Properties of Transfer Functions — 313
  - 7.6.4. Examples — 314
  - Problems — 319

Chapter 8 CIRCUIT TRANSFORMATIONS AND FILTERS
- 8.1. Frequency and Impedance Scaling — 323
- 8.2. Frequency Transformations — 326
  - 8.2.1. Prototype Networks — 326
  - 8.2.2. Low-pass to Low-pass Transformations — 327
  - 8.2.3. Low-pass to High-pass Transformations — 327
  - 8.2.4. Low-pass to Band-pass Transformations — 329
  - 8.2.5. Low-pass to Band-stop Transformations — 333
- 8.3. Impedance Transformations — 334
  - 8.3.1. On Equivalent Circuits and Impedance Transformations — 335
  - 8.3.2. Impedance Transformations in Two-port Networks — 338
  - 8.3.3. Impedance-transforming Filters — 343
- 8.4. Transfer Phase and Group Delay — 348
  - 8.4.1. Transfer Phase of a Two-port — 348
  - 8.4.2. Group Velocity and Group Delay — 350
  - 8.4.3. Group Delay of a Two-port Network — 352

CONTENTS

8.5.	All-pass Networks	353
	8.5.1. Symmetrical Lattice Networks	355
	8.5.2. The Constant Resistance Symmetrical Lattice	358
	8.5.3. The C-section	361
	8.5.4. The All-pass D-section	363
	Problems	367

Chapter 9—POSITIVE REAL IMPEDANCE FUNCTIONS

9.1.	Energy in RLC Networks	372
9.2.	Properties of Driving Point Impedances	376
	9.2.1. Consequences of Positive Real Condition	380
9.3.	The Bounded Real Condition	383
9.4.	Hurwitz Polynomials and Reactance Functions	384
9.5.	Two-Port Scattering Parameters	386
9.6.	Additional Remarks	389
	Problems	390

Chapter 10 TWO-ELEMENT-KIND SYNTHESIS

10.1.	Driving Point Impedance of LC Networks	393
	10.1.1. Some Properties of LC Impedances	395
	10.1.2. Realisation of LC Driving Point Impedances	398
	10.1.3. The Cauer Forms	404
10.2.	LC Transfer Functions	415
	10.2.1. Sufficiency of the Necessary Conditions	418
10.3.	RC Driving Point Impedances	423
10.4.	RL Driving Point Impedances	433
10.5.	Transfer Function Synthesis without Ideal Transformers	433
	10.5.1. Transmission Zeros at the Origin and Infinity	435
	10.5.2. Transmission Zeros on the Negative Real Axis	438
	10.5.3. Guillemin's Method	441
10.6.	Concluding Remarks	443
	Problems	443

Chapter 11 SYNTHESIS OF RLC DRIVING-POINT IMPEDANCES

11.1.	The Brune Procedure	449
11.2.	Darlington's Synthesis	461
11.3.	Bott and Duffin Synthesis	468
11.4.	Conclusions	475
	Problems	475

Chapter 12 TRANSFER FUNCTION SYNTHESIS

12.1.	Some General Properties	478
12.2.	Synthesis of Single-terminated Networks	481
	12.2.1. Realisation of $|Z_{21}(j\omega)|$	485
12.3.	Double-loaded Networks	486
	12.3.1. Ladder Networks	488

CONTENTS

12.4.	Approximation	492
	12.4.1. Maximally Flat Approximations	493
	12.4.2. Chebyshev Approximations	501
	12.4.3. Bandpass Approximations	511
	12.4.4. Delay Approximations	514
	12.4.5. More General Approximations	518
12.5.	Elements of Cascade Synthesis	521
12.6.	Principles of Broadband Matching Theory	532
12.7.	General Observations	542
APPENDIX		547
SOLUTIONS TO PROBLEMS		549
INDEX		575

PREFACE

This book continues the treatment of Circuit Theory begun in Volume 1 and the same basic philosophy is followed. The material presented here is suitable for the latter part of the second year and for the third year of a three years honours course in Electrical and Electronic Engineering. It has arisen from such courses given in the University of Leeds over several years. Some of the material given towards the end of Chapter 12 is included for the sake of completeness and a full treatment of this may be more appropriate to postgraduate study although the basic ideas can readily be appreciated at undergraduate level.

The material in this book consists initially of a generalisation of the analysis methods of Volume 1 and the introduction of some useful techniques of transformation. The remainder of the book is devoted to a treatment of the ideas of network synthesis. The area of linear passive networks is one of the few where exact mathematical synthesis techniques exist and it is now possible to proceed from a given specification to generate one or more networks capable of meeting these specifications. Until comparatively recently the topic of network synthesis was thought to be beyond the scope of an undergraduate course but, latterly, it has been found both possible and desirable to include such material.

The methods outlined in this book are kept deliberately quite rigorous since rigour is essential in any meaningful study of the subject of synthesis. The techniques presented here lay the foundation for an extension of the subject to other classes of networks, e.g. distributed and digital networks. Throughout these two volumes attention has been concentrated on what may now be described as the classical methods of analysis and synthesis of linear passive lumped, networks. It is, of course, quite true that in many practical situations the networks encountered may be active and/or nonlinear and in such cases some special techniques of analysis or synthesis may be neeeded. However, the basis laid here may easily be built upon to deal with such situations. As in Volume 1 we have tried to concentrate on basic principles rather than to broaden the scope in order to encompass as many classes of circuits as possible. In the usual undergraduate curriculum many opportunities to apply the ideas of circuit theory occur in other courses and the student is unlikely to be deficient in this respect.

The subject of network synthesis is sometimes thought of as being of a rather specialist nature but the ideas have wide application in for example, microwaves, solid-state device techniques and communications. It is hoped that this text will provide a suitable background for those whose prime interest is in applications to other disciplines as well as a sound foundation for those interested in further study in the area of networks.

J. O. S.
R. L.

Chapter 7

THE COMPLEX FREQUENCY PLANE

7.1. Introduction

We recall from Chapter 6 of Volume 1 that the Laplace transform of a time domain function $f(t)$ is given by

$$F(s) = \int_0^\infty f(t) \exp(-st)dt \qquad [7.1]$$

and that the Laplace transform variable $s = \sigma + j\omega$ is a complex number which is called the *complex frequency* and may be plotted in a plane called the *complex frequency plane* as shown in Fig. 7.1. Also, if the function $F(s)$ is known, the corresponding time domain function $f(t)$ may be found as the inverse Laplace transform of $F(s)$.

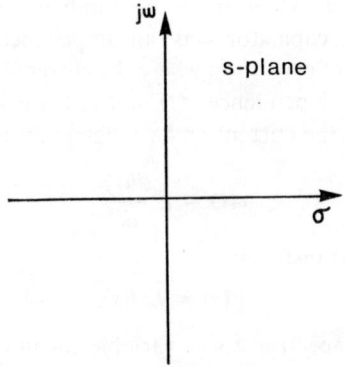

FIG. 7.1. *Complex frequency plane.*

Now if, for example, the response and excitation in some network are related through an equation such as

$$v(t) = Z(t)i(t) \qquad [7.2]$$

where $Z(t)$ is an integro-differential operator as explained in Chapter 6, then taking the Laplace transform we obtain

$$V(s) = Z(s)I(s) \qquad [7.3]$$

and a knowledge of $Z(s)$ is sufficient to calculate any response corresponding to a given excitation in terms of either the complex frequency, s, or in the time domain as seen in Section 6.7.

We shall assume in this chapter that any initial conditions are included in the form of corresponding voltage or current sources, as explained in Section 6.6, or are zero.

Our object now is to be able to express all network functions in terms of the complex frequency (or Laplace transform variable) s. This will be found to have many advantages including the ability to calculate the time domain response corresponding to any given excitation (via [7.3] and the inverse Laplace transform), and the fact that for *RLC* networks all network functions are real and rational in the variable s.

It was shown in Chapter 4 that in order to calculate network responses to steady sinusoidal excitations it is extremely convenient to consider the response to excitations of the form $\exp(j\omega t)$. This led to the use of network functions expressed as functions of ω, or more precisely as functions of $j\omega$, since every inductor has an impedance $(j\omega)L$ while every capacitor has an impedance $1/(j\omega)C$. For a more general form of excitation where the currents and voltages have some general time dependence $v(t)$ and $i(t)$, we know that the relationship between the current and voltage in an inductor is

$$v(t) = L\frac{di(t)}{dt} \qquad [7.4]$$

or taking Laplace transforms

$$V(s) = LsI(s). \qquad [7.5]$$

Thus in the Laplace transform variable an inductor acts as an impedance Ls. Similarly for a capacitor

$$i(t) = C\frac{dv(t)}{dt} \qquad [7.6]$$

or taking Laplace transforms

$$I(s) = sCV(s) \qquad [7.7]$$

so that a capacitor behaves as an impedance $1/sC$. It is therefore obvious that to calculate any network function in terms of s we merely regard the impedance of an inductor as Ls and of a capacitor as $1/Cs$ and apply the normal rules.

It is also clear that if we know some network function in terms of s, it is only necessary to replace s by $j\omega$ in order to obtain the corresponding network function appropriate to an excitation $\exp(j\omega t)$, since as seen above an inductor has an impedance sL in terms of the complex frequency variable and an impedance $(j\omega)L$ for excitations of the form $\exp(j\omega t)$, while the corresponding impedances for a capacitor are $1/sC$ and $1/(j\omega)C$ and the impedance of a resistor is the same in both cases. The process is described formally in the following way: If, say, $Z(s)$ is some impedance function written in terms of the complex frequency s, then replacing s by $j\omega$ we write $Z(j\omega)$. In general, s is a complex number but at points in the complex frequency plane of Fig. 7.1 which lie on the vertical or imaginary axis, s is purely imaginary, i.e. $\sigma = 0$ and $s = j\omega$. Thus putting $s = j\omega$ is equivalent to evaluating $Z(s)$ on the imaginary axis. The values of s which lie on the *imaginary* axis are called *real frequencies* (by convention) and when we put $s = j\omega$, we are said to evaluate the function at real frequencies and we are, in fact, finding the form of the function appropriate to an excitation of the form $\exp(j\omega t)$.

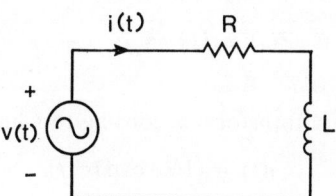

FIG. 7.2. *Series RL circuit.*

Consider, for example, the series RL circuit shown in Fig. 7.2. We know from Chapter 4 that if $v(t) = V \exp(j\omega t)$, then $i(t) = I \exp(j\omega t)$ and
$$V = I(R+j\omega L)$$
so
$$Z = V/I = R+j\omega L \qquad [7.8]$$

288   CIRCUIT THEORY

Now, in terms of the complex frequency s, we have seen that we merely treat the inductor as having an impedance sL and therefore the total impedance of the circuit is

$$Z(s) = R+sL. \qquad [7.9]$$

Also, as previously indicated, if we put $s = j\omega$, we obtain

$$Z(j\omega) = R+j\omega L \qquad [7.10]$$

which shows that $Z(j\omega)$ is the impedance for an excitation of the form $\exp(j\omega t)$. Thus, if we have a voltage $v(t)$ which is a Dirac delta function and the resulting current is $i(t)$ then

$$\mathscr{L}v(t) = 1 = V(s)$$

and therefore

$$1 = Z(s)I(s) \qquad [7.11]$$

or

$$I(s) = \frac{1}{Z(s)} = \frac{1}{R+sL} \qquad [7.12]$$

$$= \frac{1}{L[s+(R/L)]} \qquad [7.13]$$

$$= \frac{1}{L(s-s_1)}$$

so that

$$Z(s) = L(s-s_1) \qquad [7.14]$$

where $s_1 = -(R/L)$ $(=\sigma)$.

Taking the inverse transform according to Chapter 6 gives

$$i(t) = (1/L)\exp(\sigma_1 t) \qquad [7.15]$$

Thus, from a knowledge of $Z(s)$ we can readily write down the response to a sinusoidal excitation or to a delta function excitation.

7.2. The Definition of Complex Natural Frequencies

In the example given in the previous section, the particular value s_1 of the complex frequency s is obviously of considerable significance, since it occurs in both the impedance $Z(s)$ [7.14], and in the response to a delta function [7.15]. When $s = s_1$, $Z(s)$ is zero and s_1 is therefore called a *zero* of $Z(s)$. Although s_1 is real in the present example, it is regarded as a particular case of the complex frequency. s_1 is also referred to as a *natural frequency* of the network. The form of [7.15] is plotted in Fig. 7.3(a) for the cases where $\sigma < 0$ (curve (i)), $\sigma = 0$ (curve (ii)), and $\sigma > 0$ (curve (iii)). The latter case could not arise in a passive circuit since the response grows indefinitely with increasing time.

In general, the natural frequencies of a network will not necessarily be real. Imaginary ($s_1 = j\omega$) and complex ($s_1 = \sigma + j\omega$) values will also occur leading to responses of the type shown in Fig. 7.3(b), (c) and (d), when the excitation is a delta function. Such responses will be seen in more detail later. Again, a response of the type shown in Fig. 7.3(d) cannot occur in a passive network as it increases indefinitely as t tends to infinity. Thus in all cases it is clear that *the real part of a natural complex frequency must be negative or zero* in a passive network.

We shall see in later chapters that in addition to simplifying the problems of circuit analysis the ideas of complex frequency lead to a process of network synthesis whereby we may find networks which realise a specified response from a given excitation.

7.3. Poles and Zeros

In considering the series RL circuit of Fig. 7.2 we obtained the series impedance of [7.14], which leads to the definition of the zero s_1. Evidently, s_1 may be plotted as a point in the complex plane, as shown in Fig. 7.4, and this point defines the impedance

$$Z(s) = L(s-s_1)$$

completely except for the constant L, which may be regarded as a scale factor.

In general, we shall determine that any impedance or admittance function of a network is represented by the ratio of two polynomials in the complex frequency variable s. These polynomials are rational, i.e. may be factored, to give a function which, in the case of an impedance $Z(s)$, will be of the form

$$Z(s) = \frac{N(s)}{D(s)} = \frac{a_0 s^n + a_1 s^{n-1} + \ldots + a_{n-1} s + a_n}{b_0 s^m + b_1 s^{m-1} + \ldots + b_{m-1} s + b_m}$$

$$= \frac{a_0}{b_0} \frac{(s-s_1)(s-s_2)\ldots(s-s_n)}{(s-s_1')(s-s_2')\ldots(s-s_m')} \qquad [7.16]$$

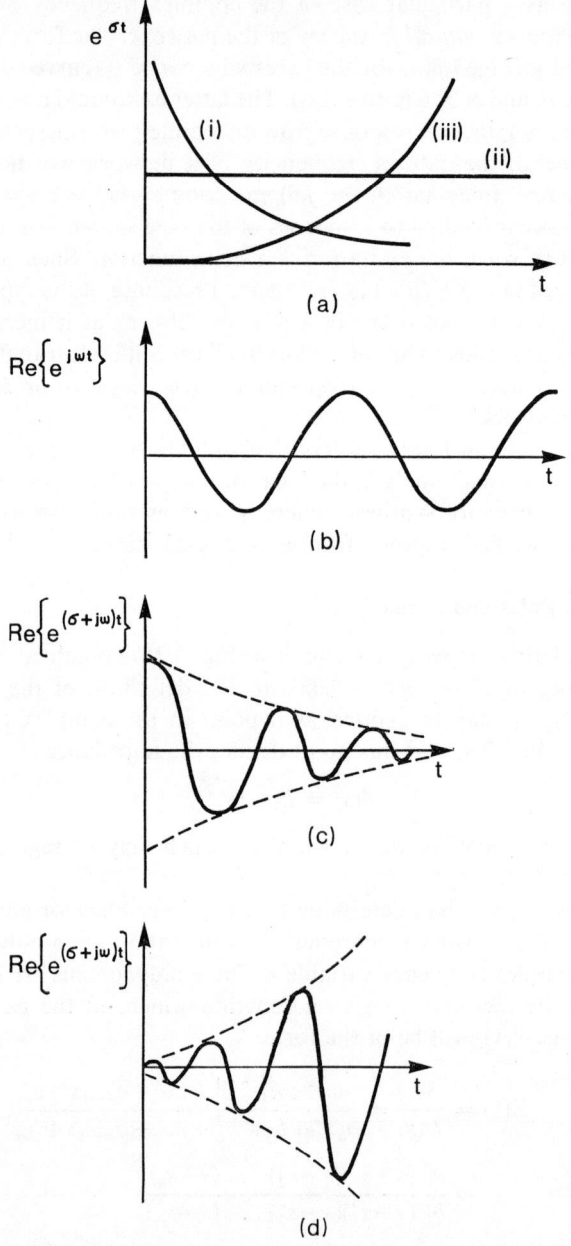

FIG. 7.3. *Representation of complex natural frequencies in the time domain* (a) $\exp(\sigma t)$ for (i) $\sigma < 0$, (ii) $\sigma = 0$, (iii) $\sigma > 0$; (b) Re $\exp(j\omega t)$; (c) Re $\exp\{(\sigma+j\omega)t\}$ for $\sigma < 0$; (d) Re $\exp\{(\sigma+j\omega)t\}$ for $\sigma > 0$.

The general proof of [*7.16*] will be given in Section 7.6. [*7.16*] is the most general form of $Z(s)$, a particular case of which for the series *RL* circuit was given by [*7.14*]. As in the latter case, the impedance $Z(s)$ in [*7.16*] is specified completely, except for a scale factor a_0/b_0, by the location in the complex plane of the *zeros* $s_1, s_2, \ldots s_n$ and of the *poles* $s'_1, s'_2 \ldots s'_m$. The term *pole* is used for a denominator root s'_i because $s = s'_i$ gives the impedance an infinite value.

7.4. Real-frequency Behaviour from Pole-zero Locations

Although we have generalised and extended the concept of frequency, the behaviour of network functions at real frequencies, i.e. for $s = j\omega$, is nearly always of primary interest. Since an impedance function $Z(s)$ is specified by the location of its poles and zeros, it is useful and interesting to investigate the real-frequency behaviour of $Z(s)$ from the pole-zero pattern.

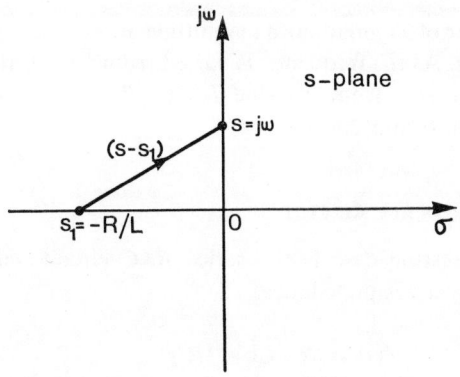

FIG. 7.4. *Representation of impedance in the complex plane.*

7.4.1. SERIES RL CIRCUIT

The first example we shall consider is the Series *RL* circuit having the single zero at $s_1 = -R/L$, as shown in Fig. 7.4. A real frequency is represented by a point on the $j\omega$ axis, and is shown as the point $j\omega$ in Fig. 7.4. Any point in the complex plane is also represented by a *vector* directed to the point from the origin, i.e. a point $s = \sigma + j\omega$ is also represented in polar form as

where
$$\left.\begin{aligned} s &= re^{j\theta} \\ r &= \sqrt{(\sigma^2 + \omega^2)} \\ \theta &= \tan^{-1} \omega/\sigma \end{aligned}\right\} \qquad [7.17]$$

i.e. a vector of modulus r and phase (or argument) θ. The vectors corresponding to the zero s_1 and the arbitrary real frequency point $j\omega$ are also shown in Fig. 7.4. We see that the vector joining s_1 to $j\omega$ is given by the difference of vectors s and s_1, which when multiplied by the constant L gives a vector equal to the impedance $Z(s)$ of [7.14] at the real frequency ω, i.e.

$$Z(s)\big|_{s=j\omega} = L(s-s_1)\big|_{s=j\omega} \qquad [7.17a]$$

The behaviour of $Z(j\omega)$ in both magnitude and phase is given directly by this vector. As the frequency is varied from 0 to ∞ the magnitude of $Z(j\omega)$ increases from a value Ls_1 ($=R$) to ∞, and the phase increases from 0 to $\pi/2$.

7.4.2. SERIES RLC CIRCUIT

A more interesting case is the series *RLC* circuit shown in Fig. 7.5(*a*), having series impedance

$$Z(s) = R + Ls + (1/Cs)$$

$$= \frac{L[s^2 + (R/L)s + (1/LC)]}{s}$$

where
$$\left.\begin{aligned} &= \frac{L(s^2 + 2\alpha s + \omega_0^2)}{s} \\ \alpha &= \frac{R}{2L}, \quad \omega_0 = \frac{1}{\sqrt{LC}} \end{aligned}\right\} \qquad [7.18]$$

i.e.

$$Z(s) = \frac{L(s-s_1)(s-s_2)}{s}$$

where

$$\left.\begin{matrix}s_1\\s_2\end{matrix}\right\} = -\alpha \pm \sqrt{(\alpha^2 - \omega_0^2)}$$

[7.19]

s_1 and s_2 are the zeros, and there is a pole of $Z(s)$ at the origin, as indicated in Fig. 7.5(b). (Zeros in the complex plane will be drawn as circles, and poles as crosses.) The zeros s_1 and s_2 are drawn for the condition $\alpha < \omega_0$, corresponding to the well-known case of a lightly-damped circuit, as discussed in Sections 3.7 and 6.4.

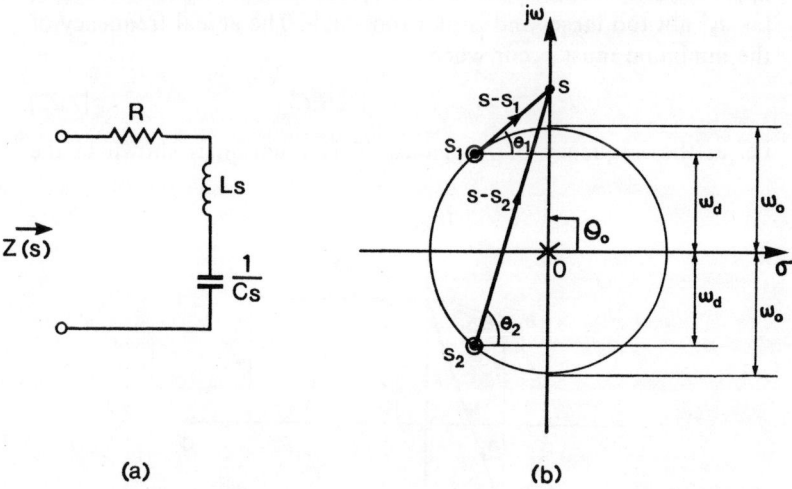

FIG. 7.5. (a) *series RLC circuit* (b) *corresponding pole-zero diagram* $[\omega_d = \sqrt{(\omega_0^2 - \alpha^2)}]$.

The real frequency behaviour, i.e. the response to a sinusoidal excitation, is given firstly by drawing the vectors from the poles and zeros to a real frequency $s = j\omega$, as shown in Fig. 7.5(b). The modulus of the impedance is then given by

$$|Z(s)| = \frac{L|s-s_1| \cdot |s-s_2|}{|s|}$$

[7.20]

i.e. the product of the zero vectors divided by the pole vector, while its phase is

$$\arg \{Z(s)\} = (\theta_1 + \theta_2) - \theta_0 \quad [7.21]$$

i.e. the difference between the sum of the arguments of the zero vectors, and the argument of the pole vector.

The behaviour of the impedance $Z(s)$ at real frequencies as ω increases from zero can be seen quite clearly by inspection of the vectors as the point s moves along the $j\omega$ axis from 0 to ∞. Thus at $\omega = 0$ the pole vector has zero length and $|Z| = \infty$, corresponding to an open circuit. Similarly at $\omega = \infty$, $|Z| = \infty$, and between these extremes of frequency there is a minimum in the value of $|Z|$. We can see that this must occur when the real frequency is located approximately opposite to the zero s_1, for then $|s-s_1|$ will be small, $|s-s_2|$ not too large, and $|s|$ not too small. The actual frequency of the minimum must occur when

$$\omega = \omega_0 = 1/\sqrt{LC} \quad [7.22]$$

i.e. at the frequency of resonance. This situation is shown in the

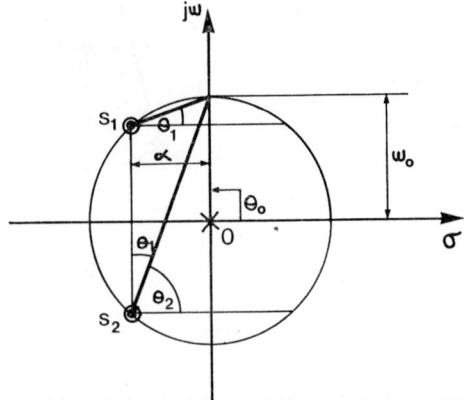

Fig. 7.6. *Pole-zero pattern of RLC circuit at resonance.*

vector diagram of Fig. 7.6, where we see that the resonant condition gives a phase of

$$\theta_1 + \theta_2 - \theta_0 = 0. \quad [7.23]$$

We can also investigate the dependence of Z on the values of s_1 and s_2. In particular it is interesting to see how the resonant value of Z changes with the location of the zeros s_1 and s_2. From [7.19] we see that for $\alpha < \omega_0$ we can write

$$\left.\begin{matrix}s_1\\s_2\end{matrix}\right\} = \sigma_1 \pm j\omega_1 = -\alpha \pm j\sqrt{(\omega_0^2 - \alpha^2)} \qquad [7.24]$$

so that

$$\sigma_1^2 + \omega_1^2 = \omega_0^2 \qquad [7.25]$$

This means that for the condition $\alpha < \omega_0$ the locus of s_1 and s_2 as ω_0 is kept constant is a semicircle of radius ω_0 whose centre is at the origin. We see that if α is small, corresponding to a small value of series resistance, there is little damping, the natural behaviour of the circuit is oscillatory, and there is a fairly sharp resonant condition. As α increases, s_1 and s_2 approach the point $(-\omega_0, 0)$. The condition when s_1 and s_2 coincide at $(-\omega_0, 0)$ corresponds to critical damping. As α increases above the value ω_0 the two zeros are real, being given by

$$\left.\begin{matrix}s_2\\s_1\end{matrix}\right\} = -\alpha \pm \sqrt{(\alpha^2 - \omega_0^2)} \qquad [7.26]$$

and we see also that the product $s_1 s_2 = \omega_0^2$. The locus of s_1 and s_2 is depicted in Fig. 7.7. The reader should compare the value of the parameters which determine whether the natural frequencies s_1 and

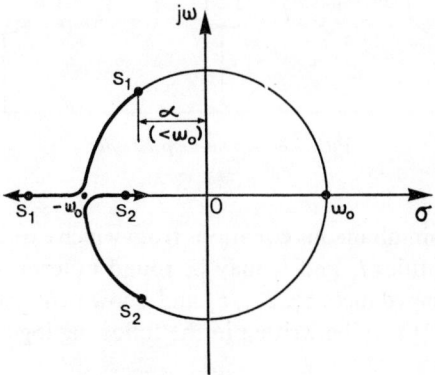

FIG. 7.7. *Locus of the zeros of an RLC circuit as the parameter α is varied with ω_0 constant.*

s_2 are complex or real with the conditions given previously in Section 3.7.

7.5. General Circuit Analysis using Complex Frequency

Kirchoff's laws were introduced in Chapter 1, and have been applied to the analysis of both d.c. and a.c. circuits (see Chapter 4 for simple examples of such analysis using complex notation). In this section we extend this analysis to the case of a general circuit containing an arbitrary number of nodes and loops. The results derived will lead to some important general theorems.

7.5.1. MESH OR LOOP ANALYSIS

Consider first a simple two-loop network as shown in Fig. 7.8. Here there are two voltage sources V_1 and V_2, and three impedances z_1, z_{12}, and z_2. Application of Kirchoff's voltage law gives

$$\left.\begin{array}{l}(z_1+z_{12})i_1-z_{12}i_2 = V_1\\ -z_{12}i_1+(z_{12}+z_2)i_2 = V_2\end{array}\right\} \quad [7.27]$$

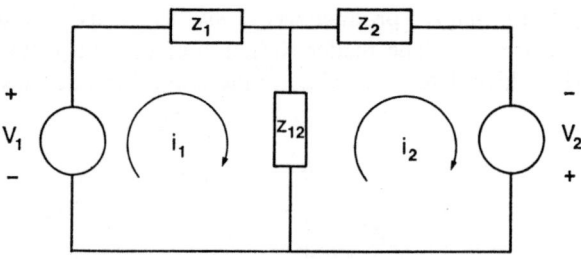

FIG. 7.8. *A two-loop circuit.*

These are two simultaneous equations from which expressions for the unknown quantities i_1 and i_2 may be found in terms of the known values of the impedances z_1, z_{12}, z_2 and known voltages V_1 and V_2. Equations [7.27] may be written in the following logical format:

$$\left.\begin{array}{l}a_{11}x_1+a_{12}x_2 = y_1\\ a_{21}x_1+a_{22}x_2 = y_2\end{array}\right\} \quad [7.28]$$

which may be solved to give

$$x_1 = \frac{a_{22}y_1 - a_{12}y_2}{a_{11}a_{22} - a_{12}a_{21}}$$

$$x_2 = \frac{-a_{21}y_1 + a_{11}y_2}{a_{11}a_{22} - a_{12}a_{21}}$$

[7.29]

Here of course we have replaced currents i_1 and i_2 by x_1 and x_2, a_{11} is $z_1 + z_{12}$, a_{22} is $z_{12} + z_2$, $a_{12} = a_{21} = -z_{12}$, and voltages V_1 and V_2 are replaced by y_1 and y_2.

Equations [7.29] can be expressed in a more compact form using determinants. The determinant of the set of Eqs. [7.28] is

$$\Delta = \begin{vmatrix} a_{11} & a_{12} \\ a_{21} & a_{22} \end{vmatrix} = a_{11}a_{22} - a_{12}a_{21} \qquad [7.30]$$

To digress a little, let us remember the definitions of the minor and cofactor of a determinant. A minor of a determinant is that determinant with one row and one column removed. Thus M_{12} is the minor obtained from the determinant Δ by removing row 1 and column 2.

A cofactor has the same definition as a minor with the exception of sign. Thus the cofactor Δ_{ij} formed by deletion of the ith row and j^{th} column of Δ is defined in terms of the minor M_{ij} by the relationship

$$\Delta_{ij} = (-1)^{i+j} M_{ij} \qquad [7.31]$$

We see that in relationship to corresponding cofactors Δ_{11} is positive, Δ_{12} negative, Δ_{13} positive, etc., i.e. cofactors progressively alternate in sign relative to the minors.

Now [7.29] may be written in determinant form as follows

$$x_1 = \frac{\Delta_{11}y_1 + \Delta_{12}y_2}{\Delta}$$

$$x_2 = \frac{\Delta_{21}y_1 + \Delta_{22}y_2}{\Delta}$$

[7.32]

where Δ is the determinant defined by [7.30].

It can be seen that the solution of [7.28] may be written down by inspection in the determinant notation of [7.32]. This form of solution is known as Cramer's rule. It is useful to extend it to a set of n simultaneous linear equations in n unknowns, i.e. the set

$$a_{11}x_1 + a_{12}x_2 + \ldots + a_{1n}x_n = y_1$$
$$a_{21}x_1 + a_{22}x_2 + \ldots + a_{2n}x_n = y_2$$
$$\ldots\ldots\ldots\ldots\ldots\ldots\ldots\ldots\ldots\ldots\ldots\ldots \quad [7.33]$$
$$\ldots\ldots\ldots\ldots\ldots\ldots\ldots\ldots\ldots\ldots\ldots\ldots$$
$$a_{n1}x_1 + a_{n2}x_2 + \ldots + a_{nn}x_n = y_n.$$

The determinant of this set of equations is

$$\Delta = \begin{vmatrix} a_{11} & a_{12} \ldots a_{1n} \\ a_{21} & a_{22} \ldots a_{2n} \\ \ldots\ldots\ldots\ldots \\ \ldots\ldots\ldots\ldots \\ a_{n1} & a_{n2} \ldots a_{nn} \end{vmatrix} \quad [7.34]$$

In order to solve [7.33] for x_1, multiply the first equation through by Δ_{11}, the second by Δ_{21}, etc., and add. The result is

$$(a_{11}\Delta_{11} + a_{21}\Delta_{21} + \ldots + a_{n1}\Delta_{n1})x_1$$
$$= \Delta_{11}y_1 + \Delta_{21}y_2 + \ldots + \Delta_{n1}y_n. \quad [7.35]$$

Notice that we have eliminated terms in $x_2, x_3, \ldots x_n$. This is because a typical coefficient of one of these, say x_2, is

$$a_{12}\Delta_{11} + a_{22}\Delta_{21} + \ldots + \Delta_{n1}a_{n1} \quad [7.36]$$

and this is the second column of [7.34] multiplied by the cofactors of the first column, which by a well-known theorem in determinant theory must be zero. To prove this, note that the coefficients in the first column do not appear in [7.36]. Hence [7.36] could be formed from [7.34] if we replace the first row by any other set of numbers, and in particular by the second row. But then [7.36] would represent the determinant of a matrix with two identical columns, which is zero. Noting also that the coefficient of x_1 in [7.35] is, by definition, the value of the determinant [7.34], we obtain

$$x_1 = \Delta_{11}y_1 + \Delta_{21}y_2 + \ldots + \Delta_{n1}y_n$$

and similarly

$$\left.\begin{aligned} x_2 &= \Delta_{12}y_1 + \Delta_{22}y_2 + \ldots + \Delta_{n2}y_n \\ &\qquad\ldots\ldots\ldots\ldots\ldots\ldots\ldots\ldots\ldots \\ &\qquad\ldots\ldots\ldots\ldots\ldots\ldots\ldots\ldots\ldots \\ x_n &= \Delta_{1n}y_1 + \Delta_{2n}y_2 + \ldots + \Delta_{nn}y_n \end{aligned}\right\} \quad [7.37]$$

Yet another and even simpler way to solve [7.33] is by use of matrix algebra. [7.33] may be written in the form

$$\begin{bmatrix} a_{11} & a_{12} \ldots a_{1n} \\ a_{21} & a_{22} \ldots a_{2n} \\ \ldots\ldots\ldots\ldots \\ \ldots\ldots\ldots\ldots \\ a_{n1} & a_{n2} \ldots a_{nn} \end{bmatrix} \begin{bmatrix} x_1 \\ x_2 \\ .. \\ .. \\ x_n \end{bmatrix} = \begin{bmatrix} y_1 \\ y_2 \\ .. \\ .. \\ y_n \end{bmatrix} \quad [7.38]$$

or

$$\mathbf{ax} = \mathbf{y} \quad [7.39]$$

where **x** and **y** are column matrices (or vectors) and **a** is a square matrix.

The solution of [7.39] may be written down immediately as

$$\mathbf{x} = \mathbf{a}^{-1}\mathbf{y} \quad [7.40]$$

or in long form as

$$\begin{bmatrix} x_1 \\ x_2 \\ .. \\ .. \\ x_n \end{bmatrix} = 1/\Delta \begin{bmatrix} \Delta_{11} & \Delta_{21} \ldots \Delta_{n1} \\ \Delta_{12} & \Delta_{22} \ldots \Delta_{n2} \\ \ldots\ldots\ldots\ldots \\ \ldots\ldots\ldots\ldots \\ \Delta_{1n} & \Delta_{2n} \ldots \Delta_{nn} \end{bmatrix} \begin{bmatrix} y_1 \\ y_2 \\ .. \\ .. \\ y_n \end{bmatrix} \quad [7.41]$$

For the proof of the form of [7.41] consult any text on matrix theory for the definition and proof of the inverse of a square matrix. The reader should also be able to supply his own proof by using the argument used earlier to derive [7.35].

300   CIRCUIT THEORY

These results are invaluable in the solution of the sets of simultaneous linear equations formed when applying Kirchoff's laws to the solution of linear network problems. Thus returning to our problem of Fig. 7.8 wherein Kirchoff's voltage laws are expressed as [7.27], we can write down the solution for the unknown currents i_1 and i_2 by Cramer's rule. The determinant for the equations is

$$\Delta = \begin{vmatrix} z_1+z_{12} & -z_{12} \\ -z_{12} & z_2+z_{12} \end{vmatrix} = (z_1+z_{12})(z_2+z_{12})-z_{12}^2 \quad [7.42]$$

and the cofactors are

$$\Delta_{11} = z_2+z_{12} \quad \Delta_{12} = z_{12}$$
$$\Delta_{21} = z_{12} \quad \Delta_{22} = z_1+z_{12} \quad [7.43]$$

so that the result for i_1 and i_2 is

$$i_1 = \frac{z_2+z_{12}}{\Delta} V_1 + \frac{z_{12}}{\Delta} V_2$$

$$i_2 = \frac{z_{12}}{\Delta} V_1 + \frac{z_1+z_{12}}{\Delta} V_2 \quad [7.44]$$

Now let us extend this result to the general circuit case where we have n loops as shown in Fig. 7.9. The impedances in the circuit are defined as follows:

Let the total impedance common to loops j and k be z_{jk}.
Let the total impedance in any loop j which is *not* shared by any other loop be z_j.

Hence the sum of all impedances in loop j is

$$z_{jj} = z_j+z_{j1}+z_{j2}+\ldots z_{jk}+\ldots+z_{jn} \quad [7.45]$$

where $k \neq j$. z_{jj} is defined as the self impedance of loop j. z_{jk} is defined as the mutual impedance common to loops j and k. For example in the case of the two-loop circuit of Fig. 7.8 the self impedances of the two loops are z_1+z_{12} and z_2+z_{12}, while the mutual impedance is z_{12}.

Returning to the n-loop network, let us define by V_j the sum total voltage in loop j due to voltage sources which may be in that loop. In Fig. 7.9 voltage sources of V_1 and V_2 are shown in loops 1 and 2. Note also the sign convention to be used in the loop equations given

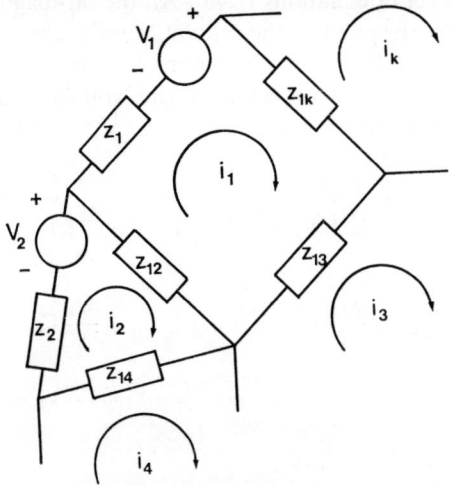

FIG. 7.9. *Part of a general* n-*loop circuit.*

below. All loop currents are defined in a clockwise direction, and V_j takes a positive value on the right hand side of the equations if the + sign corresponds to the emergence of current. Kirchoff's equations may now be written down in the form

$$z_{11}i_1 - z_{12}i_2 - z_{13}i_3 \ldots - z_{1n}i_n = V_1$$
$$-z_{21}i_1 + z_{22}i_2 - \ldots \ldots - z_{2n}i_2 = V_2$$
$$\ldots \ldots \ldots \ldots \ldots \ldots \ldots \ldots \ldots \ldots \ldots \ldots \quad [7.46]$$
$$\ldots \ldots \ldots \ldots \ldots \ldots \ldots \ldots \ldots \ldots \ldots \ldots$$
$$-z_{n1}i_1 - z_{n2}i_2 \ldots \ldots \ldots + z_{nn}i_n = V_n$$

In this set of equations the only terms on the left hand side which have a positive sign are those of the form z_{jj} as given by [7.45], i.e. the self impedance terms. These are terms on the main diagonal of the *loop impedance matrix*

$$[z] = \begin{bmatrix} z_{11} & -z_{12} & -z_{13} \ldots -z_{1n} \\ -z_{21} & z_{22} & -z_{23} \ldots -z_{2n} \\ \ldots \ldots \ldots \ldots \ldots \ldots \ldots \ldots \\ -z_{n1} & -z_{n2} & \ldots \ldots \ldots z_{nn} \end{bmatrix} \quad [7.47]$$

formed by the set of equations [7.46]. All the off-diagonal terms are negative, and correspond to the mutual impedances.

The solution of [7.46] for the unknown currents $i_1, i_2, \ldots i_n$ in terms of all the known impedances and voltage sources may be written down immediately using matrix algebra—see [7.33] to [7.41]. The result is

$$i_1 = \frac{\Delta_{11}}{\Delta} V_1 + \frac{\Delta_{21}}{\Delta} V_2 + \ldots + \frac{\Delta_{n1}}{\Delta} V_n$$

$$i_2 = \frac{\Delta_{12}}{\Delta} V_1 + \frac{\Delta_{22}}{\Delta} V_2 + \ldots + \frac{\Delta_{n2}}{\Delta} V_n$$

$$\ldots\ldots\ldots\ldots\ldots\ldots\ldots\ldots\ldots\ldots\ldots\ldots \quad [7.48]$$

$$i_n = \frac{\Delta_{1n}}{\Delta} V_1 + \frac{\Delta_{n2}}{\Delta} V_2 + \ldots + \frac{\Delta_{nn}}{\Delta} V_n$$

where Δ is the determinant of matrix [7.47] and the Δ_{jk} are the various cofactors. It is important to take note of the following definitions of notation in the various equations:

(a) The self and mutual impedances are denoted by lower case z. Upper case Z will be used later to define another type of impedance for the same general network.

(b) The circulating currents are defined by lower case i. Upper case I will be used later to define an ideal current source.

(c) The ideal voltage sources are defined by upper case V. Lower case v will be used later in nodal analysis to define a nodal voltage.

The question may arise as to how one would analyse the general network if it were to contain ideal current sources in addition to ideal voltage sources. The answer is that if we decide to use loop analysis we would first convert the ideal current sources into ideal voltage sources using the source transformations described in Sections 1.4.2 and 6.6.1. However we could alternatively use nodal analysis, our next topic, in which case we would transform any voltage sources present into ideal current sources. The choice as to whether loop or nodal analysis should be used is based mainly on the desirability of having the least number of equations. The topic was covered in Section 1.5. Occasionally, however, a particular type of analysis may be preferred because of its more obvious and natural ease of application, e.g. if the circuit contains only current sources one would use nodal analysis.

7.5.2. NODAL ANALYSIS

This is the dual of loop analysis and utilises Kirchoff's current law which states that the sum of all currents entering a point in a network

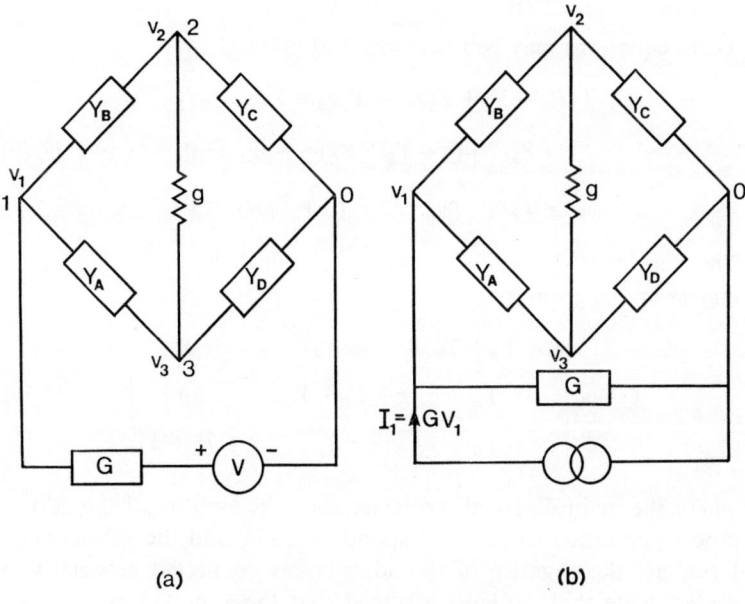

FIG. 7.10. *Wheatstone bridge.*

is zero. Such a point is defined as a node. As an example of nodal analysis consider the Wheatstone bridge network shown in Fig. 7.10(a). This network contains four nodes. One of these is chosen as the reference node and arbitrarily given a zero voltage. The voltages at the other three nodes are v_1, v_2, and v_3.

The first step in the analysis is to replace the ideal voltage source V in series with the admittance G by the ideal current source I in parallel with the admittance G, giving the network of Fig. 7.10(*b*).

The current entering node 1 is I_1, and the current leaving is Gv_1 directly to node 0, $Y_B(v_1 - v_2)$ to node 2, and $Y_A(v_1 - v_3)$ to node 3. The current balance equation is therefore

$$Gv_1 + Y_B(v_1 - v_2) + Y_A(v_1 - v_3) = I_1$$

Similarly the equations for the other two nodes are

$$Y_B(v_2 - v_1) + g(v_2 - v_3) + Y_C v_2 = 0$$

$$Y_A(v_3 - v_1) + g(v_3 - v_2) + Y_D v_3 = 0$$

These equations may be re-arranged to give

$$(G + Y_A + Y_B)v_1 - Y_B v_2 - Y_A v_3 = I_1$$
$$-Y_B v_1 + (g + Y_B + Y_C)v_2 - g v_3 = 0 \quad [7.49]$$
$$-Y_A v_1 - g v_2 + (g + Y_A + Y_D)v_3 = 0$$

The matrix of this set of equations is called the *nodal admittance matrix* and is given by

$$(y) = \begin{vmatrix} G + Y_A + Y_B & -Y_B & -Y_A \\ -Y_B & g + Y_B + Y_C & -g \\ -Y_A & -g & g + Y_A + Y_D \end{vmatrix} \quad [7.50]$$

where the main diagonal terms are each the sum of all the admittances connected to the corresponding node, and the off-diagonal terms are the negative of the admittances connected between two nodes. Note that we have assumed that the admittances have the same values for either direction of current passing through them, so that the entry y_{jk} in [7.50] is equal to the entry y_{kj}. This may not always be the case, e.g. if one of the admittances were a diode. All the admittances in Fig. 7.10 are the same independent of the direction of the current, and are known as *bilateral* elements. Any network containing only bilateral elements is defined as a *reciprocal* network. It is characterised by the fact that the *jk* and corresponding *kj*

THE COMPLEX FREQUENCY PLANE

entries of either its loop impedance or nodal admittance matrix representations are equal.

Continuing with our nodal analysis theory, it is now quite simple to see how to analyse the general $(n+1)$ node network shown in Fig. 7.11. Node $n+1$ is chosen as the reference having zero (or ground) potential, and the other n nodes are assumed to have

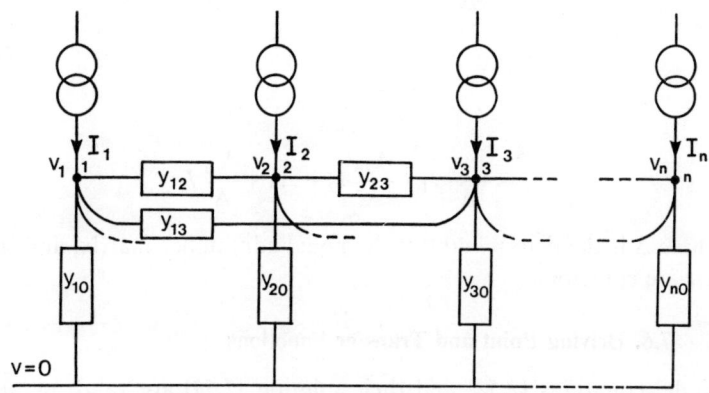

FIG. 7.11. *General network for nodal analysis.*

unknown potentials $v_1, v_2, \ldots v_n$. The admittance from the jth node to ground is y_{j0}, and the *mutual admittance* between modes j and k is y_{jk}. The *self admittance* of the jth node is given by

$$y_{jj} = y_{j0} + y_{j1} + \ldots + y_{j,j-1} + y_{j,j+1} + \ldots + y_{jn} \qquad [7.51]$$

Finally current sources $I_1, I_2, \ldots I_n$ are connected to each node (except ground). The nodal analysis equations may be written down immediately in the form

$$y_{11}v_1 - y_{12}v_2 - y_{13}v_3 \ldots -y_{1n}v_n = I_1$$
$$-y_{21}v_1 + y_{22}v_2 - y_{23}v_3 \ldots -y_{2n}v_n = I_2$$
$$\ldots \qquad [7.52]$$
$$-y_{n1}v_1 - y_{n2}v_2 \ldots \ldots + y_{nn}v_n = I_n$$

These equations are the duals of those given in [7.46], and may be solved similarly using matrix theory. The result is

$$v_1 = \frac{\Delta_{11}}{\Delta}I_1 + \frac{\Delta_{21}}{\Delta}I_2 + \ldots + \frac{\Delta_{n1}}{\Delta}I_n$$

$$v_2 = \frac{\Delta_{12}}{\Delta}I_1 + \frac{\Delta_{22}}{\Delta}I_2 + \ldots + \frac{\Delta_{n2}}{\Delta}I_n$$

$$\ldots\ldots\ldots\ldots\ldots\ldots\ldots\ldots\ldots\ldots\ldots\ldots\ldots$$ [7.53]

$$\ldots\ldots\ldots\ldots\ldots\ldots\ldots\ldots\ldots\ldots\ldots\ldots\ldots$$

$$v_n = \frac{\Delta_{1n}}{\Delta}I_1 + \frac{\Delta_{2n}}{\Delta}I_2 + \ldots + \frac{\Delta_{nn}}{\Delta}I_n$$

where Δ is the determinant of the nodal admittance matrix, and the Δ_{jk} are cofactors.

7.6. Driving Point and Transfer Functions

Nodal equations [7.52] and their solution [7.53] are valid for any linear network and for any type of current sources which drive the network. Now consider the case where the network is driven by only one current source, say I_1. Then [7.53] simplifies considerably to give

$$v_k = \frac{\Delta_{1k}}{\Delta}I_1$$

$$k = 1, 2, \ldots n \qquad [7.54]$$

Hence the impedance looking into port 1 formed by terminal 1 and ground is

$$Z_{11} = \frac{v_1}{I_1}\bigg|_{I_2 = I_3 = \ldots = I_n = 0}$$

$$= \frac{\Delta_{11}}{\Delta} \qquad [7.55]$$

Z_{11} is known as the *open-circuited driving-point impedance*, and is a generalisation of that defined in Section 5.2 for a two port. It is the impedance looking into the 'driven' port 1 when all other current sources in the network are open circuited.

THE COMPLEX FREQUENCY PLANE

Now consider port 1 and any other port, say port 2, for which we define

$$Z_{21} = \frac{v_2}{I_1}\bigg|_{I_2 = I_3 = \ldots = I_n = 0} = \frac{\Delta_{12}}{\Delta} \qquad [7.56]$$

as the *open circuited transfer impedance* between ports 1 and 2.

Similar definitions for the dual quantities may be obtained by commencing from the loop equations [7.46] and their solution in the form of [7.48]. In the case where only voltage source v_1 is present, all other sources being shorted, we see that [7.48] reduces to

$$i_k = \frac{\Delta'_{1k}}{\Delta'} V_1$$

$$k = 1, 2, \ldots n \qquad [7.57]$$

The primes introduced into [7.57] serve only to distinguish the various determinants and cofactors from those of [7.54], which are not related.

The admittance looking into port 1 with all voltage sources short-circuited other than the first (which drives the rest of the network) is

$$Y_{11} = \frac{i_1}{V_1}\bigg|_{V_2 = V_3 = \ldots = V_n = 0} = \frac{\Delta'_{11}}{\Delta'} \qquad [7.58]$$

Y_{11} is defined as the *short circuited driving point admittance* of the network.

The *short circuited transfer admittance* between loops 1 and 2 is defined as

$$V_{21} = \frac{i_2}{V_1}\bigg|_{V_2 = V_3 = \ldots = V_n = 0} = \frac{\Delta'_{12}}{\Delta'} \qquad [7.59]$$

It is important not to confuse the driving point and transfer immittances defined in this section with the self and mutual immittances comprising the impedance and admittance matrices [7.47] and [7.52]. The latter are denoted by lower case letters, and are actual immittances (impedances or admittances) forming the network. The driving point quantities on the other hand, denoted by upper case letters, are immittances seen looking into the entire network, and in general would involve all elements in the network. Similarly the transfer immittances are also formed by the ratio of

B

two determinants, involving all components in the network, in general a very complicated expression.

As already stated, the various cofactors of the loop impedance and nodal admittance matrices of the same network are not directly related to one another. For example the two matrices may even be of different order, for the number of loop equations is not necessarily equal to the number of nodal equations. Hence the transfer quantities are certainly not reciprocally related, i.e.

$$Z_{jk} \neq \frac{1}{Y_{jk}} \quad (j \neq k) \qquad [7.60]$$

Having made this point we now proceed to demonstrate one relationship which is valid, namely that the driving point impedance (calculated from the nodal admittance matrix) and the driving point

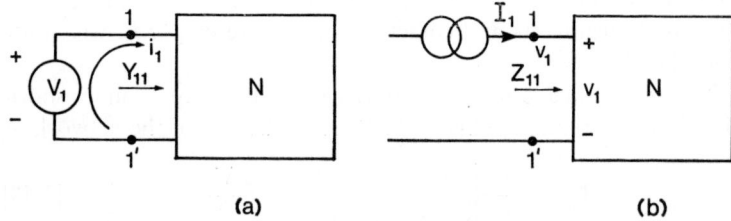

(a)   (b)

FIG. 7.12. *Relationship between methods of calculating a driving point immittance.*

admittance (calculated from the loop impedance matrix) for the same port of the same network are reciprocally related. We consider a network N having a pair of accessible terminals $11'$ (or port) as shown in Fig. 7.12(a) and (b). Figure 7.12(a) shows this port being driven by a voltage source and Fig. 7.12(b) by a current source. Now in calculating Z_{11} from the nodal admittance matrix all internal generators in the network are reduced to zero and similarly in calculating Y_{11} from the loop impedance matrix all internal generators are reduced to zero. Thus the measurement at the port $11'$ in either Fig. 7.12(a) or (b) refers to exactly the same conditions in the network N and since N is linear it does not matter whether we measure by applying a current source or a voltage source. Thus

$$Z_{11} = 1/Y_{11} \qquad [7.61]$$

Driving point and transfer functions are important quantities in circuit theory and in practical network applications, since, as we have seen in Chapter 5 they tell us how a circuit 'looks' at an accessible port or how it will behave if used to transfer energy.

7.6.1. PROPERTIES OF DRIVING POINT AND TRANSFER FUNCTIONS

Let us now consider the special case where the driving source is sinusoidal, i.e. in the case of a current source it will be represented as $I_1 e^{j\omega t}$ at port 1. All nodal voltages in the network are forced to take the form $v_k e^{j\omega t}$, and the $e^{j\omega t}$ factors will cancel out of the nodal equations represented by [7.52].

It is always possible to choose the nodes so that the self and mutual admittances take the general form

$$y_{jk} = G_{jk} + j\omega C_{jk} + \frac{1}{j\omega L_{jk}} \qquad [7.62]$$

i.e. the most complicated form of [7.62] consists of a conductance, a capacitance, and an inductance in parallel. If more complicated forms occur then one or more extra nodes may be introduced to simplify the expression to the simple [7.62] type. We are now in a position to make the further simplification introduced in Section 7.1, wherein we replace $j\omega$ by the complex frequency variable s, giving

$$y_{jk} = G_{jk} + sC_{jk} + \frac{1}{sL_{jk}} \qquad [7.63]$$

which is now a *real* function of a *complex* variable s. Moreover y_{jk} is a simple rational function consisting of the ratio of two polynomials, a numerator of second degree and the denominator of first degree, i.e.

$$y_{jk} = \frac{C_{jk}s^2 + G_{jk}s + 1}{L_{jk}s} \qquad [7.64]$$

In Section 7.3 we stated a general form for impedance or admittance functions of networks as the ratio of two rational polynomials in s, and stated [7.16]. This form follows immediately from our general expressions for driving point and transfer immittances. For example the driving point impedance formed from the admittance matrix is given in [7.55] as $Z_{11} = \Delta_{11}/\Delta$ where all entries in

Δ_{11} and Δ take the form of [7.64]. Hence the result for each of the determinants Δ and Δ_{11} is the sum of various products of simple rational functions like [7.64]. The result must take the form of [7.16], which is the ratio of two rational polynomials in s with real coefficients. By the fundamental theorem of algebra, an nth degree polynomial has n roots, so that any driving point or transfer function may be expressed in factored form as

$$Z_{jk} = \frac{a_0(s-s_1)(s-s_2)\ldots(s-s_n)}{b_0(s-s'_1)(s-s'_2)\ldots(s-s'_m)} \qquad [7.65]$$

We recall that the zeros $s_1, s_2, \ldots s_n$ and poles $s'_1, s'_2, \ldots s'_m$ are natural frequencies of the network.

Since we know that the numerator and denominator polynomials in [7.65] have only real coefficients, this tells us that the poles and zeros must be either real or occur as complex conjugate pairs. For, supposing either polynomial were to contain a pair of factors

$$(s+a_1+jb_1)(s+a_2+jb_2) \qquad [7.66]$$

neither root of which may be paired with its complex conjugate root elsewhere. [7.66] may be expanded to give the quadratic in s:

$$(s+a_1)(s+a_2)-b_1b_2+j[b_1(s+a_2)+b_2(s+a_1)]$$

Since the overall polynomial must contain only real coefficients, the imaginary component of the quadratic must vanish, i.e.

$$b_1(s+a_2)+b_2(s+a_1) = 0$$

This identity must be true for any value of the variable s, so that

$$b_1+b_2 = 0; b_1a_2+b_2a_1 = 0$$

i.e.

$$b_2 = -b_1; \quad a_2 = a_1$$

So that the second factor in [7.66] is actually $(s+a_1-jb_1)$. The root of this factor is the complex conjugate of that of the first factor, proving that any complex root must occur in conjunction with its complex conjugate. This is true for both driving point and transfer functions.

7.6.2. SPECIAL PROPERTIES OF DRIVING POINT FUNCTIONS

Returning to our general circuit and consideration of driving point functions, Fig. 7.13(a) shows the port driven by a current source I_1, and Fig. 7.13(b) shows the Laplace transformed network with the current source given its frequency domain representation $I_1(s)$. The dual configuration was shown in Fig. 6.9. Again, in the previous Chapter, Section 6.8, the significance of the network response

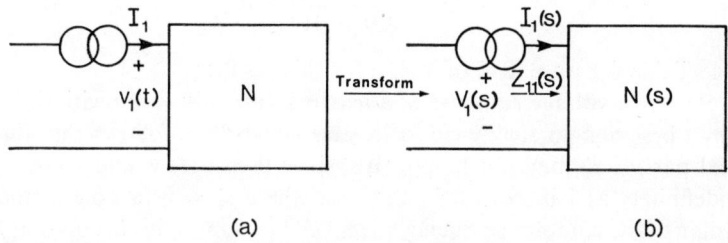

FIG. 7.13. *Time and frequency domain relationships.*

obtained when I_1 is a Dirac delta function was explained. Here we have the dual result, that the driving point impedance is the transform of the voltage response to a unit delta-function current source, i.e.

$$Z_{11}(s) = \frac{V_1(s)}{I_1(s)} = V_1(s)$$

$$(I_1(s) = 1) \quad [7.67]$$

so that

$$v_1(t) = \mathscr{L}^{-1} Z_{11}(s) \quad [7.68]$$

The driving point impedance $Z_{11}(s)$ has the form of [7.65], the ratio of two rational polynomials with real coefficients, shown in factored form. In order to find the voltage response to the delta-function current source we must take the inverse Laplace transform. Thus it is necessary first to express [7.65] as a partial fraction expansion (see Section 6.4.4). In general we would have to assume that some of the poles are zero or multiple, or both zero and multiple. Thus a pole factor $(s-s'_1)$ of order j would contribute a series to the partial fraction expansion of the form

$$\sum_{1}^{j} \frac{A_r}{(s-s'_1)^r} \qquad [7.69]$$

and in general the complete partial fraction expansion of $Z_{11}(s)$ would consist of the sum of several series like [7.69].

The inverse Laplace transform of [7.69] is given as

$$v_j(t) = \sum_{1}^{j} \frac{A_r}{(r-1)!} t^{r-1} e^{s'_1 t} \qquad [7.70]$$

(see Table 6.1 on p. 236 of Vol. 1 for this result).

Since the voltage response of a network can only *decay* with time, or at best remain steady (for a lossless network) it follows that the real part of s'_1 may not be positive, since then $v_j(t)$ would increase indefinitely as t increases. In the case where $s'_1 = 0$ (a pole at the origin of the complex frequency plane) $e^{s'_1 t} = 1$, and this means that j can not be greater than 1, else we would find $v_j(t)$ increasing to infinity as t tends to infinity. In cases where s'_1 is purely imaginary giving a pole on the real frequency axis then again the multiplicity can not be greater than 1. In the general case where s'_1 is located in the left half (real part negative) s-plane but not on the $j\omega$ axis then any multiplicity is allowable. This statement follows from the general result

$$\lim_{t \to \infty} t^{r-1} e^{\sigma'_1 t} = 0 \qquad [7.71]$$

where σ'_1 is negative.

The above results for the driving point impedance $Z_{11}(s)$ relate only to the poles s_j in the denominator of [7.65]. However we have seen that the driving point impedance and admittance are the reciprocals of one another [7.61] so that the poles of $Y_{11}(s)$ are the zeros of $Z_{11}(s)$ and vice versa. Now we can carry out the dual argument for $Y_{11}(s)$ and form the current time domain response to a delta-function voltage driving source as

$$i_1(t) = \mathscr{L}^{-1} Y_{11}(s) \qquad [7.72]$$

Hence by repeating the preceding arguments we can show that all the properties of the poles of $Z_{11}(s)$ hold also for the zeros since these are the poles of $Y_{11}(s)$.

THE COMPLEX FREQUENCY PLANE

In summary these properties are as follows:
1. The poles and zeros of the driving point functions are either real, or occur in complex conjugate pairs.
2. They all occur in the left half s-plane or on the $j\omega$ axis.
3. Poles and zeros on the $j\omega$ axis are simple, i.e. can never have multiplicity (or degree) greater than unity.

From result (3) the following property follows as a corollary:
4. The degrees of the numerator and denominator polynomials of the driving point functions can not differ by more than one.

For if they do, there will be a pole (at $s = \infty$) of degree at least two in the driving point impedance or admittance, which is not allowed.

7.6.3. SPECIAL PROPERTIES OF TRANSFER-FUNCTIONS

The restrictions on transfer functions are not so restrictive as those on driving point functions. This may be seen from previous work, namely that the transfer function is the ratio of two different quantities in the network, e.g. [7.56] and [7.59], and that the transfer quantities are not reciprocally related, e.g. [7.60]. Figure 7.14 depicts

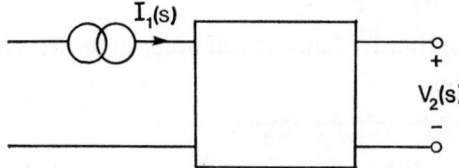

FIG. 7.14. *Definition of $Z_{21}(s)$*.

a network driven by the transformed current source $I_1(s)$ at port 1 resulting in the voltage $V_2(s)$ at port 2. The transfer impedance is

$$Z_{21}(s) = \frac{V_2(s)}{I_1(s)} \qquad [7.73]$$

and if the current source is actually a Dirac delta function, this leads to the voltage at port 2 in the time domain given as

$$v_2(t) = \mathscr{L}^{-1} Z_{21}(s) \qquad [7.74]$$

314   CIRCUIT THEORY

which is similar to [7.72]. We may now apply the same reasoning to $v_2(t)$ as we used in the previous section in respect of the driving point voltage, and we obtain the same results, namely that the poles of $Z_{21}(s)$ lie only in the left half s-plane or on the real frequency axis, where they must be simple. However we can not form a quantity $Y_{21}(s)$ reciprocal to $Z_{21}(s)$ which has any physical significance, and hence we are not restricted in any way in the choice of the zeros of $Z_{21}(s)$. We can say that if

$$Z_{21}(s) = \frac{N_{21}(s)}{D_{21}(s)} \qquad [7.75]$$

then the degree of the denominator $D_{21}(s)$ may be greater than the degree of $N_{21}(s)$ by, perhaps, more than unity, while the degree of $N_{21}(s)$ may not be greater than that of $D_{21}(s)$ by more than unity else the inverse Laplace transform would become infinite at infinite time.

These results are summarized below, and should be compared with the results given for driving point functions in the previous section.

1. The poles and zeros of transfer functions are either real, or occur in complex conjugate pairs.

2. The poles of transfer functions all occur in the left half s-plane or on the $j\omega$ axis.

3. Poles on the $j\omega$ axis are simple.

4. The degree of the numerator may not exceed that of the denominator by more than one.

7.6.4. EXAMPLES

The following examples serve to illustrate many of the results relating to driving point and transfer functions. The first is the two-port network shown in Fig. 7.15, where we are required to find the driving point impedance at port 1 when port 2 is open circuited, and the transfer impedance from port 1 to port 2 under the same condition.

There is more than one way to tackle this problem. We have been making extensive use of the nodal analysis method in this chapter, and so we shall demonstrate this method first. There are five distinct nodes in the circuit, one is chosen as the datum, and the others labelled 1 to 4 as illustrated. Actually here port 2 is associated with

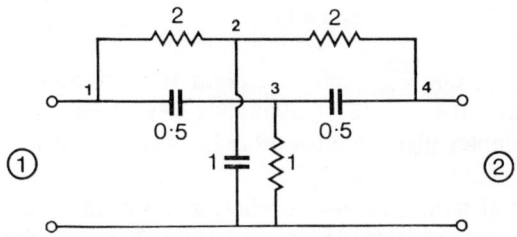

FIG. 7.15. *Example 1. Element values are Ohms and Farads.*

node 4, but the nodes may be in any order. The nodal admittance matrix may be written down by inspection as

$$(y) = \begin{bmatrix} \frac{1}{2}(1+s) & -\frac{1}{2} & -\frac{1}{2}s & 0 \\ -\frac{1}{2} & 1+s & 0 & -\frac{1}{2} \\ -\frac{1}{2}s & 0 & 1+s & -\frac{1}{2}s \\ 0 & -\frac{1}{2} & -\frac{1}{2}s & \frac{1}{2}(1+s) \end{bmatrix} \qquad [7.76]$$

and the required driving point and transfer impedances are

$$Z_{11} = \frac{\Delta_{11}}{\Delta} \; ; \qquad Z_{41} = \frac{\Delta_{41}}{\Delta} \qquad [7.77]$$

The determinant Δ is simplified by subtracting column 4 from column 1, and then adding row 1 to row 4, giving

$$\Delta = \begin{bmatrix} \frac{1}{2}(1+s) & -\frac{1}{2} & -\frac{1}{2}s & 0 \\ 0 & 1+s & 0 & -\frac{1}{2}s \\ 0 & 0 & 1+s & -\frac{1}{2}(1+s) \\ 0 & -1 & -s & \frac{1}{2}(1+s) \end{bmatrix}$$

$$= \tfrac{1}{2}s(s+1)^2$$

It is simple to find Δ_{11} and Δ_{41} directly and the results are

$$\Delta_{11} = (s+1)(s^2+4s+1)/4$$
$$\Delta_{41} = (s+1)(s^2+1)/4$$

so that the driving point and transfer impedances are

$$Z_{11} = \frac{s^2+4s+1}{2s(s+1)}; \quad Z_{41} = \frac{s^2+1}{2s(s+1)} \qquad [7.78]$$

Note that Z_{11} and Z_{41} each have poles at $s = 0$ and $s = -1$, and Z_{11} has zeros at $s = -(2+\sqrt{3})$ and $-(2-\sqrt{3})$, which are all in the left half complex plane. On the other hand the transfer function Z_{41} has a pair of zeros at $s = \pm j$.

At the real frequency $\omega = 1$ where $s = j$ we note that $Z_{41} = 0$, which means that at this frequency there is no oscillation at the output terminals. This is a physical interpretation of a *transmission zero*.

Although the nodal analysis method is direct it is better to avoid using it whenever possible since it is quite easy to make a mistake when evaluating a determinant such as [7.76]. A more elegant method to solve the problem is to note that the circuit of Fig. 7.15 consists of two circuits in parallel, and to form the two port admittance matrix by adding the individual matrices of these two circuits. Since the two circuits are each T networks we first write down their impedance matrices (see Section 5.6) which for the T containing the two series resistors and the shunt capacitor is

$$(z_1) = \begin{bmatrix} 2+1/s & 1/s \\ 1/s & 2+1/s \end{bmatrix}$$

and for the T containing the two series capacitors and the shunt resistor is

$$(z_2) = \begin{bmatrix} 1+2/s & 1 \\ 1 & 1+2/s \end{bmatrix}$$

These are converted into admittance matrices using the results given in Section 5.7, leading to

$$(y_1) = \frac{s}{4(s+1)} \begin{bmatrix} 2+1/s & -1/s \\ -1/s & 2+1/s \end{bmatrix}$$

$$(y_2) = \frac{s^2}{4(s+1)} \begin{bmatrix} 1+2/s & -1 \\ -1 & 1+2/s \end{bmatrix}$$

Hence the overall admittance matrix is

$$(y) = (y_1)+(y_2) = \frac{1}{4(s+1)}\begin{bmatrix} s^2+4s+1 & -(s^2+1) \\ -(s^2+1) & s^2+4s+1 \end{bmatrix}$$

In order to find the open circuit impedance parameters the impedance matrix must be formed. The determinant of (y) is

$$\det y = \frac{(s^2+4s+1)^2-(s^2+1)^2}{16(s+1)^2}$$

$$= \frac{(2s^2+4s+2)(4s)}{16(s+1)^2} = \frac{s}{2}$$

Hence the open circuit driving point impedance is

$$z_{11} = \frac{y_{22}}{\det y} = \frac{s^2+4s+1}{2s(s+1)}$$

and the open circuit transfer impedance is

$$z_{21} = \frac{-y_{21}}{\det y} = \frac{s^2+1}{2s(s+1)}$$

in agreement with the results obtained in [7.78] by nodal analysis.

The second example shown in Fig. 7.16 is rather simpler, but uses different analysis techniques. Again the question is to find the

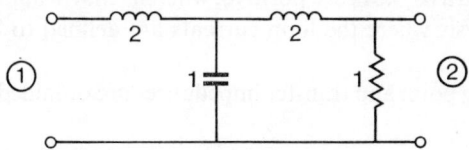

FIG. 7.16. *Example 2. Element values are Ohms, Farads and Henrys.*

driving point impedance Z_{11} and the transfer impedance Z_{21} with port 2 open-circuited. A simple analysis method is to use the transfer matrix technique, multiplying transfer matrices of the individual components, as outlined in Section 5.12. In this instance we can regard the 1 ohm shunt resistor as a termination at port 2, and multiply the transfer matrices of the other three components to give

$$\begin{bmatrix} 1 & 2s \\ 0 & 1 \end{bmatrix} \begin{bmatrix} 1 & 0 \\ s & 1 \end{bmatrix} \begin{bmatrix} 1 & 2s \\ 0 & 1 \end{bmatrix} = \begin{bmatrix} 2s^2+1 & 4s^3+4s \\ s & 2s^2+1 \end{bmatrix}$$

The driving point impedance is obtained by application of [5.61], i.e.

$$Z_{11} = \frac{AZ+B}{CZ+D} = \frac{(2s^2+1).1+4s^3+4s}{s.1+2s^2+1}$$

$$= \frac{4s^3+2s^2+4s+1}{2s^2+s+1} \quad [7.79]$$

while the transfer impedance Z_{21} is obtained from [5.64] as

$$Z_{21} = \frac{Z}{CZ+D} = \frac{1}{2s^2+s+1} \quad [7.80]$$

Alternatively we can write down the impedance matrix of the first three components using [5.27] as

$$(z) = \begin{bmatrix} 2s+1/s & 1/s \\ 1/s & 2s+1/s \end{bmatrix} \quad [7.81]$$

It is worth emphasizing here that this two port impedance matrix is defined with the currents at the two ports defined as positive into the network, as in Fig. 5.6. This explains why the terms in the z_{12} and z_{21} position in matrix [7.81] are positive, whereas they would be negative in loop analysis where the loop currents are defined to be all in the same direction.

The driving point and transfer impedances are obtained from [5.42] and [5.43], i.e.

$$Z_{11} = \frac{\det z + z_{11}Z}{Z+z_{22}}$$

$$= \frac{(2s+1/s)^2 - 1/s^2 + (2s+1/s)}{1+2s+1/s} = \frac{4s^3+2s^2+4s+1}{2s^2+s+1}$$

$$Z_{21} = \frac{z_{21}Z}{Z+z_{22}} = \frac{1/s}{1+2s+1/s} = \frac{1}{2s^2+s+1}$$

in agreement with [7.80] and [7.81] obtained by the use of the transfer matrix.

In the latter part of this chapter we have introduced some general properties of driving point and transfer functions. These will be extended and elaborated on in Chapter 9 with a view to the development of *synthesis* techniques for networks. Prior to this in the next chapter we will be dealing with some rather interesting types of networks and network transformation techniques.

Problems

7.1. Find the driving point admittance of the circuit shown in Fig. 7.17. Plot the locations of the poles and zeros in the complex s-plane for the following two cases

(a) $L = C = R = 1$, $r = 0.1$.

(b) $L = C = R = r = 1$.

Deduce the general behaviour of the driving point admittance at real frequencies from the pole-zero locations in the two cases. Estimate the peak value of the driving point admittance, and give rough sketches of the magnitude of $Y(s)$ in each case.

Which is the more useful circuit for rejection of the frequency $\omega = 1$ from the resistor R?

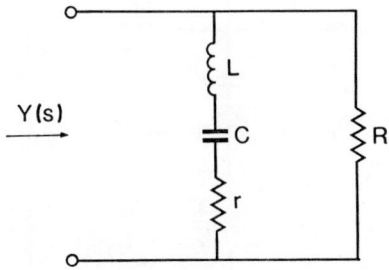

Fig. 7.17. *Problem 7.1.*

7.2. Find the driving point impedance of the network shown in Fig. 7.18 and plot the poles and zeros in the *s*-plane. Draw a sketch of the behaviour of the modulus of the driving point impedance.

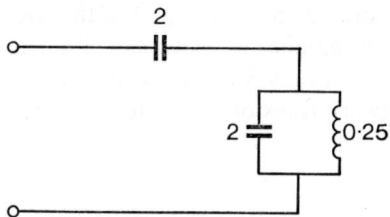

Fig. 7.18. Problem 7.2. Element values are Ohms, Farads, Henrys.

7.3. Add a resistance $R = 2$ in parallel with the inductor of Fig. 7.18, and find the new driving point impedance and *s*-plane pole-zero pattern. What is the essential difference between the networks before and after the resistance is added?

7.4. Write down the loop analysis equations for the three port network shown in Fig. 7.19. Find the driving point impedance at port 1 and the transfer impedances between ports 1 and 2 and also between ports 1 and 3. Find the square of the modulus of these transfer functions at real frequencies, and comment on the results.

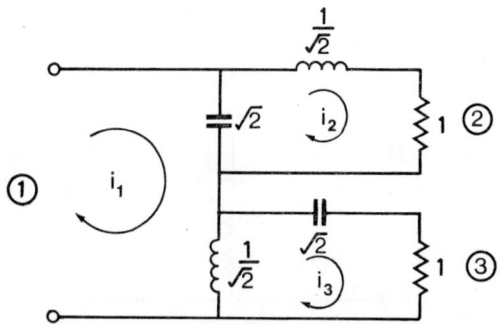

Fig. 7.19. Problem 7.4. Element values are Ohms, Farads, Henrys.

7.5. Given a passive linear network of $(n+1)$ nodes each of which is connected to a current generator, write down the nodal analysis equations. Define the self and mutual admittances. Show that these admittances are rational functions of the complex frequency variable s.

7.6. Define the natural frequencies of a network, and describe the restrictions on the driving point and transfer functions. Verify these restrictions for the network shown in Fig. 7.20.

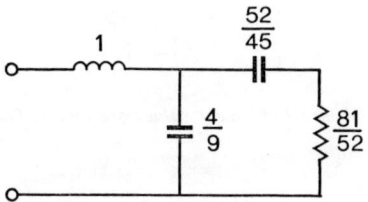

Fig. 7.20. Problem 7.6. Element values are Ohms, Farads, Henrys.

7.7. (a) Fig. 7.21 shows two capacitively coupled *Pi* networks forming a four port network. Find the relationship between the voltages and currents at the four ports.

(b) Find expressions for the driving point impedance Z_{11} and the transfer impedance Z_{14} when ports 2 and 3 are open circuited (i.e. $I_2 = I_4 = 0$) and port 3 is terminated by a unity resistance.

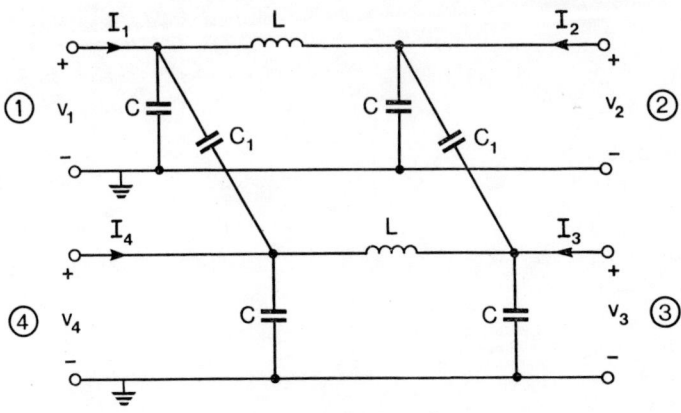

Fig. 7.21. Problem 7.7.

7.8. Find the driving point impedance of the accessible port of the network shown in Fig. 7.22.

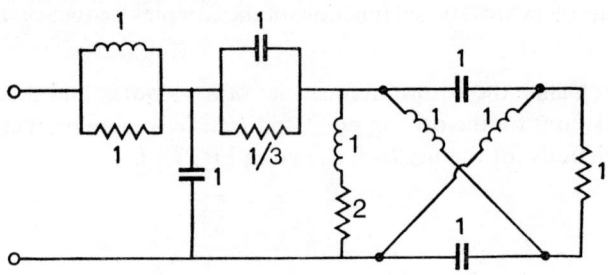

FIG. 7.22. Problem 7.8. Element values are *Ohms, Farads, Henrys*.

Chapter 8

CIRCUIT TRANSFORMATION AND FILTERS

8.1. Frequency and Impedance Scaling

The reader may have noticed that throughout this book, whenever we have treated examples we have been using circuit element values and frequencies of the order of unity, e.g., 1 Farad, 1 Henry, 1 Ohm, and 1 radian per second. Of course, values such as these are quite impractical and are rarely, if ever, encountered in real life, but their use does make the arithmetical calculations considerably less complicated. This use of values close to unity is to be regarded as a practice of standardisation, or more commonly referred to as *normalisation*. One reason why it is so useful is that it is very easy to convert a *normalised* network to be a practical network by a process of *denormalisation*.

Since we assume that we are dealing with *linear* networks denormalisation is very simple, especially in the case of the impedance and was discussed under the heading of 'scaling of networks' in Section 4.13. If we have designed the network to work at an 'impedance level' of 1Ω, and now wish to convert it to a level of $R_1\Omega$, we simply multiply all impedances in the network by R_1. The impedance level of 1Ω may be the value of a terminating resistance at one or more ports of the network, and this becomes $R_1\Omega$ after denormalisation. An inductance L becomes an inductance R_1L, and a capacitance C becomes a capacitance C/R_1. Any ideal transformers in the network remain unchanged.

Frequency scaling is equally straightforward. Supposing we have designed a network to operate at a frequency $\omega = 1$ and we now require the network to have the same behaviour at the frequency $\omega = \omega_1$; then all circuit elements are required to have that impedance at ω_1 which they had previously at $\omega = 1$. Since resistances and ideal transformers are frequency independent they remain unchanged, but reactive circuit elements must be changed. Considering an inductance L which changes to a value L_1, we know that to obtain the same impedance at the frequency ω_1 we must have

$$L = \omega_1 L_1 \qquad [8.1]$$

Similarly a capacitance C has admittance jC at $\omega = 1$ and changes to a value $j\omega C_1$ at $\omega = \omega_1$, giving

$$C_1 = C/\omega_1 \qquad [8.2]$$

Now we may combine the processes of impedance and frequency scaling into one statement. In order to multiply the impedance level by R_1 and multiply the frequency by ω_1, we multiply all resistances by R_1, all inductances by R_1/ω_1, and all capacitances by $1/R_1\omega_1$.

Before proceeding with an illustrative example of such scaling it is interesting to establish the effect of frequency scaling on the time domain behavior of the network. The magnitude scaling, has no effect on the time domain behavior since the natural frequencies of the network are not affected by a change of impedance level. Actually it is intuitively obvious that if we scale the frequency by a factor ω_1 the time constants in the network must reduce by the same factor. In this case the proof is quite simple if we recall the Laplace transform integral relationship

$$F(s) = \int_0^\infty f(t) e^{-st}\, dt \qquad [8.3]$$

which relates the frequency and time domain responses of network functions $F(s)$ and $f(t)$ respectively. The scaling relationship we require is given in [6.18], i.e.

$$\mathscr{L}\{f(at)\} = \frac{1}{a} F\left(\frac{s}{a}\right) \qquad [8.4]$$

If we scale the time axis by the factor a by introducing the new time scale

$$t' = at \qquad [8.5]$$

then [8.4] becomes

$$\mathscr{L}^{-1} F(s/a) = af(t') \qquad [8.6]$$

[8.6] states that if we scale the axes of the s-plane by the factor a, then the time domain response is changed in time scale by the factor a according to [8.5] and also in time response magnitude by the same factor. For example, if $a = 10$, this is equivalent to scaling the frequency by a factor 10, and the time domain response is speeded up by the factor 10 while the amplitude of this response is increased by

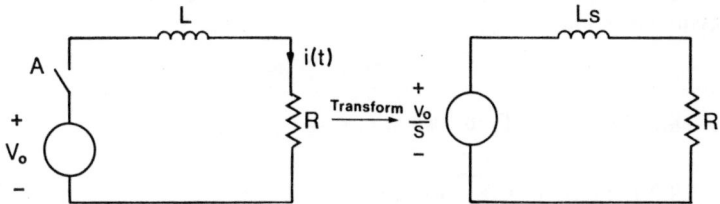

FIG. 8.1. *Series RL circuit.*

the factor 10. To take a more specific example, consider the series RL circuit shown in Fig. 8.1, excited by closing the switch A at time $t = 0$. The current $I(s)$ is given as

$$I(s) = \frac{V_0}{s(R+Ls)} \qquad [8.7]$$

leading to the time domain current response

$$i(t) = \frac{V_0}{R}[1-\exp\{(-R/L)t\}] \qquad [8.8]$$

Now let us find out the effect of scaling s by a factor a, i.e.

$$I\left(\frac{s}{a}\right) = \frac{V_0}{(s/a)[R+L(s/a)]}$$

$$= \frac{aV_0}{R}\left(\frac{1}{s} - \frac{1}{s+Ra/L}\right) \qquad [8.9]$$

This gives a time domain response

$$i'(t) = \frac{aV_0}{R}[1-\exp\{(-R/L)at\}]$$

$$= ai(at) \qquad [8.10]$$

where we have utilised [8.8] to obtain [8.10], which is seen to be in agreement with [8.6]. The finite natural frequency of [8.10] is given as

$$s_1 = -aR/L \qquad [8.11]$$

326   CIRCUIT THEORY

which has increased by the factor a compared with the unscaled network. We see that in practice the result can be obtained by dividing the value of the inductance by a and multiplying the source voltage by a. The response time is speeded up by the factor a (assumed >1).

8.2. Frequency Transformations

8.2.1. PROTOTYPE NETWORKS

At the start of this chapter we mentioned the fact that it is often convenient to use normalised circuit element values and normalised frequency. Another important reason why we study normalised networks is that such networks are the natural result of applying the network synthesis techniques we shall introduce later in this volume. For example, if we wish to design a two-port band pass filter network operating between given resistive terminations, the most commonly used synthesis technique is first to design a two-port low pass filter operating between 1Ω resistances and having a cut-off frequency of 1 rad/sec. This is then transformed into the required band pass filter using various frequency and impedance transformations. The low pass filter is referred to as a *prototype* network since it is the prototype from which many other types of networks may be derived. In addition to conversion to a band pass filter, the prototype may also be readily converted into a high-pass or band-stop filter. This process of using prototype networks reduces the amount of work required in network design since we can compile tables of prototypes which may be used to design a wide variety of different networks.

A typical prototype network having three reactive circuit-elements is shown in Fig. 8.2(*a*), and its insertion loss is typically of the form shown in Fig. 8.2(*b*). The same example with the element values $C_1 = 1$, $L_2 = 2$, $C_3 = 1$, is given in Fig. 5.23. It is convenient to introduce a single symbol g_r ($r = 1, 2, 3$) to represent the capacitances and inductances, as shown in Fig. 8.2(*a*). Note that g_r is the susceptance of a shunt element or the reactance of a series element at the frequency $\omega = 1$, which is the cut-off frequency of the low-pass filter. In terms of the complex frequency variable s the immittance (impedance or admittance as appropriate) of the rth element is $g_r s$.

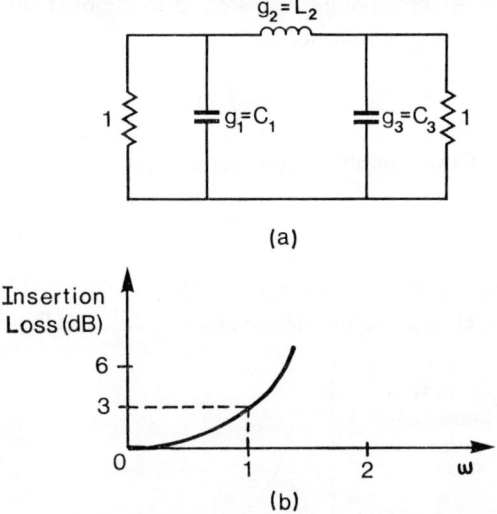

FIG. 8.2. (a) *Low-pass prototype network;* (b) *Typical response for such a network.*

8.2.2. LOW-PASS TO LOW-PASS TRANSFORMATIONS

If we wish to retain the low-pass characteristic of the prototype but change the cut-off frequency from $\omega = 1$ to $\omega = \omega'$, we merely apply the frequency transformation discussed earlier in this chapter. The use of the 'g' notation enables [*8.1*] and [*8.2*] to be combined to read

$$g'_r = g_r/\omega' \qquad [8.12]$$

where the g'_r are the new L or C values in the transformed low-pass filter.

8.2.3. LOW-PASS TO HIGH-PASS TRANSFORMATIONS

The insertion loss characteristic of a high-pass filter is shown in Fig. 8.3(*a*). The filter has a low insertion loss at all frequencies higher than the cut-off frequency, which in the figure is normalised to $\omega = 1$. The attenuation is high below $\omega = 1$. This characteristic is a kind of opposite to that of the low-pass filter shown in Fig. 8.2(*b*), with the pass and stop bands being interchanged. In fact it may be seen that one characteristic is transferred to the other by means of a reciprocal transformation, i.e.

$$\omega \to 1/\omega$$

or in terms of the complex frequency variable [8.13]

$$s \to 1/s$$

The high-pass filter resulting from [8.13] is shown in Fig. 8.3(b). In obtaining the element values therein, for example the first inductance,

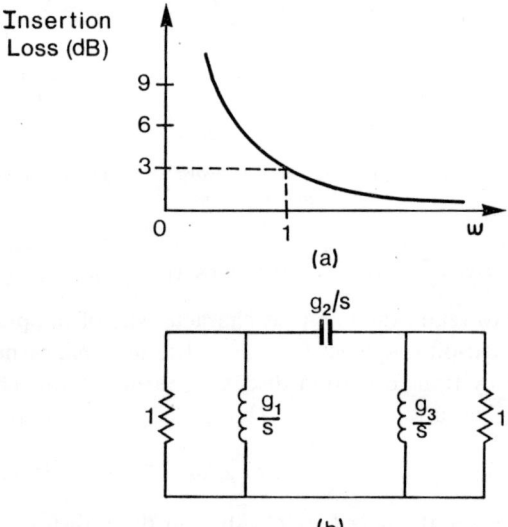

FIG. 8.3. (a) *Typical insertion loss of a high-pass filter;* (b) *Corresponding circuit derived from Fig. 8.2(a).*

note that the first element of the prototype of Fig. 8.2(a) has admittance $g_1 s$, and by virtue of [8.13] this is transformed to admittance g_1/s, or a shunt impedance of s/g_1. Hence this is represented by an inductance of value $1/g_1$, as shown. It is seen that the high-pass filter is the *dual* of the low-pass filter.

In order to transform the cut-off frequency to some other value $\omega \neq \omega'$ it is necessary only to apply [8.12], which applies to both low-

and high pass networks (or to any other characteristic for that matter).

8.2.4. LOW-PASS TO BAND-PASS TRANSFORMATIONS

The insertion loss characteristic of a band pass filter derived from the prototype of Fig. 8.2(a) must take the form shown in Fig. 8.4(a).

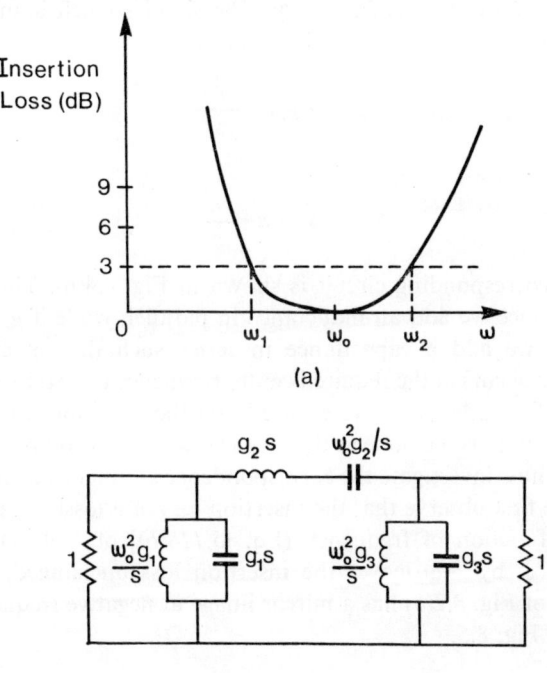

FIG. 8.4. (a) *Band-pass insertion loss characteristic;* (b) *Corresponding circuit derived from Fig. 8.2(a).*

The mid band or band centre frequency ω_0 corresponds to $\omega = 0$ of the prototype, while both band edge frequencies correspond to $\omega = 1$ of the prototype. Frequencies $\omega = 0$ and $\omega = \infty$ correspond to $\omega = \infty$ of the prototype.

It is possible to derive the required frequency transformation from this information but it is more instructive to derive it from physical rather than purely mathematical reasoning. In order to reproduce the zero insertion loss of the low pass filter (which occurs at $\omega = 0$) at the frequency $\omega = \omega_0$ in the band pass filter it is necessary to resonate each element of the low pass filter. In the low pass filter zero insertion loss occurs when each immittance ($g_r s$) is zero. Thus in the band pass case we require immittances which are zero at the centre frequency $s = j\omega_0$. The values of the prototype immittances at this frequency are $j\omega_0 g_r$ so we must add to each an immittance whose value is $-j\omega_0 g_r$ at this frequency. The simplest such immittance is $\omega_0^2 g_r / s$ so that the transformation required is

$$g_r s \to g_r s + \frac{\omega_0^2 g_r}{s} \qquad [8.14]$$

or

$$s \to s + \frac{\omega_0^2}{s}$$

and the corresponding circuit is shown in Fig. 8.4(b). Thus if g_r is a capacitance we add an inductance in parallel, while if g_r is an inductance we add a capacitance in series such that in each case resonance occurs at the required centre frequency ω_0. So by means of this transformation we have ensured that the behaviour of the band pass filter at ω_0 is the same as that of the low pass prototype at $\omega = 0$. We must now investigate the correspondence at other frequencies. To do so, we first observe that the insertion loss of a lossless network is an even function of frequency (Eq. [5.118(b)] of Vol. 1) so that replacing ω by $-\omega$ leaves the insertion loss unchanged, and the response of Fig. 8.2(b) has a mirror image at negative frequencies as shown in Fig. 8.5.

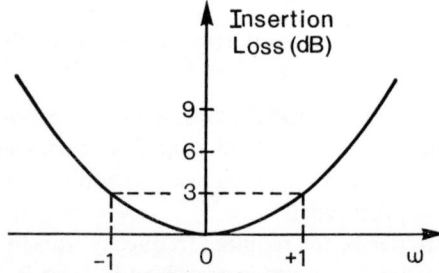

FIG. 8.5. *Low-pass insertion loss characteristic plotted for positive and negative frequencies.*

It is clear from Fig. 8.4(*a*) that there are two positive frequencies corresponding to each value of insertion loss (as well as two negative values in accordance with the observation made above). The question we ask is as follows: If a particular insertion loss L_1, say, occurs at frequency Ω (and of course $-\Omega$) in the low pass prototype response, at what frequencies (say ω_A and ω_B) does it occur in the bandpass response? Obviously, if all immittances of the low pass prototype at a particular frequency are equal to the corresponding immittances of the band pass filter at a corresponding frequency then the insertion losses are equal. Thus for the insertion loss L_1 the immittances of the low pass prototype are $\pm jg_r\Omega$ and corresponding immittances of the band pass response must have the same value. Thus we put

$$\pm jg_r\Omega = jg_r[\omega - (\omega_0^2/\omega)]$$

and ω_A and ω_B are the two positive roots of these equations hence

$$\omega - (\omega_0^2/\omega) = \pm\Omega$$

and the four roots are given by

$$\omega = \tfrac{1}{2}(\pm\Omega \pm \sqrt{(\Omega^2 + 4\omega_0^2)})$$

The positive roots (assuming $\omega_A > \omega_B$) are therefore

$$\omega_A = \tfrac{1}{2}(\sqrt{(\Omega^2 + 4\omega_0^2)} + \Omega)$$
$$\omega_B = \tfrac{1}{2}(\sqrt{(\Omega^2 + 4\omega_0^2)} - \Omega)$$

The interesting properties of this solution are that

$$\omega_A \omega_B = \omega_0^2 \qquad [8.15]$$

and

$$\omega_A - \omega_B = \Omega \qquad [8.16]$$

Thus any two frequencies having the same attenuation are such that their product is the square of the centre frequency and the response is described as being *geometrically symmetrical* about the centre frequency. It is interesting to note that the simple tuned circuits introduced in Section 4.11 have this property which is illustrated in Section 7.4.2. The second property is also of major importance. It states that the difference between any two (positive) frequencies having the same attenuation is equal to the (positive) frequency in the low pass prototype at which this attenuation occurs. Thus, for example, in the response of Fig. 8.2(*b*) the attenuation is 3dB at

$\omega = 1$, so that in the band pass response of Fig. 8.4(a) the difference between the frequencies where the attenuation is 3dB, $(\omega_2 - \omega_1)$ is also unity. This transformation is therefore said to conserve bandwidth.

If we require a band pass filter whose bandwidth (difference between frequencies corresponding to some specified attenuation) is different from that of the low pass prototype we merely frequency scale the prototype appropriately. Thus if we require a 3dB bandwidth $\omega_2 - \omega_1$ which is not unity we first frequency scale

$$g_r \to \frac{g_r}{\omega_2 - \omega_1}$$

and then carry out the low pass-band pass transformation on the scaled prototype i.e.

$$\frac{g_r s}{\omega_2 - \omega_1} \to \frac{g_r}{\omega_2 - \omega_1}\left(s + \frac{\omega_0^2}{s}\right)$$

or combining both transformations we get the general transformation

$$g_r s \to \frac{g_r}{\omega_2 - \omega_1}\left(s + \frac{\omega_0^2}{s}\right)$$

from the low pass response of bandwidth unity to a band pass response, centre frequency ω_0 of bandwidth $\omega_2 - \omega_1$. The transformation may also be written

$$s \to \frac{\omega_0}{\omega_2 - \omega_1}\left(\frac{s}{\omega_0} + \frac{\omega_0}{s}\right)$$

$$= Q\left(\frac{s}{\omega_0} + \frac{\omega_0}{s}\right) \qquad [8.17]$$

where

$$Q = \frac{\omega_0}{\omega_2 - \omega_1} \qquad [8.18]$$

is called the loaded Q-factor of the response and its reciprocal

$$w = \frac{1}{Q} = \frac{\omega_2 - \omega_1}{\omega_0} \qquad [8.19]$$

is called the *fractional bandwidth* or sometimes the decrement.

8.2.5. LOW-PASS TO BAND-STOP TRANSFORMATIONS

The object of a band stop filter is to present high attenuation in a restricted frequency band while passing all other frequencies. Such a characteristic derived from the low pass prototype of Fig. 8.2 or 8.5

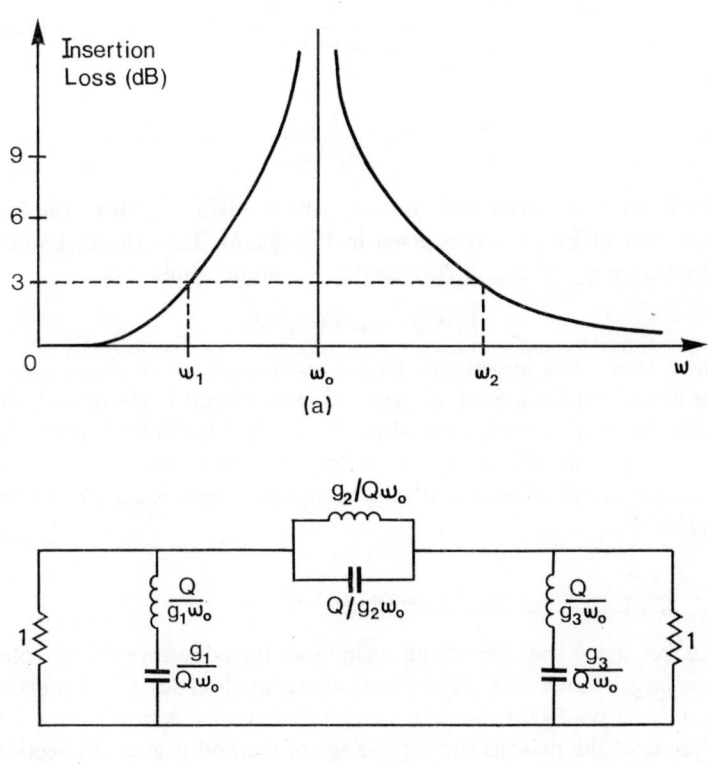

FIG. 8.6. (a) *Bandstop insertion loss characteristic;* (b) *Corresponding circuit resulting from the application of* [8.20] *to Fig. 8.2(a).*

is shown in Fig. 8.6(*a*), the cut-off frequencies $\omega = \pm 1$ of the prototype are transferred to the frequencies ω_1 and ω_2 in the band-stop filter, the frequency $\omega = 0$ of the prototype transforms into the frequencies 0 and ∞ in the band stop filter, and the frequency $\omega = \infty$ of the prototype transforms to ω_0. The frequency transformation required is a combination of the low to high pass and low to band pass transformations, for it can be seen that the band stop filter may be formed by transforming the high pass filter depicted in Fig. 8.3 by means of the transformation given by [8.17]. Combining [8.13] and [8.17] we derive the low-pass to band-stop transformation

$$s \to \frac{1}{Q[(s/\omega_0)+(\omega_0/s)]} \qquad [8.20]$$

where

$$Q = \frac{\omega_0}{\omega_2 - \omega_1}$$

The band-stop filter circuit obtained by applying [8.20] to the low pass filter of Fig. 8.2(a) is given in Fig. 8.6(b). Thus the first shunt admittance $g_1 s$ of Fig. 8.2(a) becomes a shunt admittance,

$$g_1/Q[(s/\omega_0)+(\omega_0/s)],$$

which is a series impedance $Q[(s/g_1\omega_0)+(\omega_0/g_1 s)]$ in shunt across the input terminals. Such a series resonant circuit in shunt with the main filter is also known as a *shunt Foster section*. Similarly the series element $g_2 s$ of the low-pass prototype transforms into a shunt resonant circuit in series with the main line, which is a *series Foster section*.

8.3. Impedance Transformations

The frequency transformation techniques introduced in this chapter have largely replaced the image parameter method described in many earlier text books for the design of filter networks. A brief discussion of some of the reasons for the change of method is given in Section 5.14.2. To this point we are able to design two-port filters operating between equal resistive terminations.

Sometimes it is necessary to design a filter to operate between unequal resistances. This is known as an impedance-transforming filter. The simplest way to derive the impedance transformation is by incorporation into the network of the ideal transformer of Fig. 5.19. However in practice it is often impossible to build such a device, especially at either very low or very high frequencies (below 1 KHz or above 500 MHz), because of the large physical size of the transformer, its finite insertion loss, and/or parasitic effects, in other words because of its lack of perfection.

The study of transformerless networks has been of considerable interest in circuit theory. The term 'transformerless' means a network designed without using mutual inductances. It does not mean that impedance transformation may not take place, as we shall discover in the next section.

8.3.1. ON EQUIVALENT CIRCUITS AND IMPEDANCE TRANSFORMATIONS

A very effective method of impedance transformation is to scale corresponding rows and columns of the admittance or impedance matrices by a convenient constant. The method may be illustrated in general by considering such an operation on the circuit of Fig. 7.11 which possesses the nodal admittance formulation given in [7.52], i.e.

$$\begin{bmatrix} y_{11} & -y_{12} \cdots -y_{1n} \\ -y_{21} & y_{22} \cdots -y_{2n} \\ \cdots\cdots\cdots\cdots\cdots\cdots \\ \cdots\cdots\cdots\cdots\cdots\cdots \\ -y_{n1} & -y_{n2} \cdots \ y_{nn} \end{bmatrix} \begin{bmatrix} v_1 \\ v_2 \\ \cdot \\ \cdot \\ v_n \end{bmatrix} = \begin{bmatrix} I_1 \\ I_2 \\ \cdot \\ \cdot \\ I_n \end{bmatrix} \qquad [8.21]$$

Now let us carry out the following arbitrary mathematical transformation on the admittance matrix of [8.21], namely multiply the first row and first column of the matrix by the constant m. As far as the network is concerned the element values and hence the nodal voltages will change, but the driving currents we will maintain at their original values. The new nodal admittance formulation is

$$\begin{bmatrix} m^2 y_{11} & -m y_{12} & -m y_{13} \cdots -m y_{1n} \\ -m y_{21} & y_{22} & -y_{23} \cdots -y_{2n} \\ \cdot & \cdot & \cdot \\ \cdot & \cdot & \cdot \\ -m y_{n1} & -y_{n2} & -y_{n3} \cdots\cdots y_{nn} \end{bmatrix} \begin{bmatrix} v'_1 \\ v'_2 \\ \cdot \\ \cdot \\ v'_n \end{bmatrix} \begin{bmatrix} I_1 \\ I_2 \\ \cdot \\ \cdot \\ I_n \end{bmatrix}$$

[8.22]

where the new nodal voltages are distinguished from the voltages in the original unscaled network by the prime symbols. The solution of [8.22] for the new nodal voltages is

$$
\left.\begin{array}{l}
v'_1 = \dfrac{\Delta'_{11}}{\Delta'} I_1 + \dfrac{\Delta'_{21}}{\Delta'} I_2 + \ldots + \dfrac{\Delta'_{n1}}{\Delta'} I_n \\[1em]
v'_2 = \dfrac{\Delta'_{12}}{\Delta'} I_1 + \dfrac{\Delta'_{22}}{\Delta'} I_2 + \ldots + \dfrac{\Delta'_{n2}}{\Delta'} I_n \\[1em]
\quad \cdot \qquad \cdot \qquad \cdot \\
\quad \cdot \qquad \cdot \qquad \cdot \\
\quad \cdot \qquad \cdot \qquad \cdot \\
v'_n \quad \dfrac{\Delta'_{1n}}{\Delta'} I_1 + \dfrac{\Delta'_{2n}}{\Delta'} I_2 + \ldots + \dfrac{\Delta'_{nn}}{\Delta'} I_n
\end{array}\right\} \qquad [8.23]
$$

where Δ' is the determinant of the scaled admittance matrix of [8.22] and Δ'_{kj} are cofactors. However by direct comparison between the original and scaled admittance matrices it is elementary to show that

$$\Delta' = m^2 \Delta \qquad [8.24]$$

for multiplication of any row or column of a determinant by m multiplies the determinant by m; and here we have multiplied one row and then one column each by m. Similarly we can show the following relationship

$$\left.\begin{array}{l} \Delta'_{11} = \Delta_{11} \\ \Delta'_{1k} = m\Delta_{1k} \\ \Delta'_{jk} = m^2 \Delta_{jk} \quad (j, k \neq 1) \end{array}\right\} \qquad [8.25]$$

Recalling the definition of the driving point impedance as [7.55] we see that the scaled driving point impedance at port 1 is given as

$$Z'_{11} = \frac{\Delta'_{11}}{\Delta'} = \frac{1}{m^2} \frac{\Delta_{11}}{\Delta} = \frac{Z_{11}}{m^2} \qquad [8.26]$$

Hence the driving point impedance at port 1 is reduced by the factor m^2, in other words we have effectively introduced an ideal transformer with a turns ratio of $1:m$ at that port.

It is interesting to see the effect of this transformation on the other network quantities. For example the transfer impedance between ports 1 and k is given in [7.56] as

$$Z'_{1k} = \frac{\Delta'_{1k}}{\Delta'} = \frac{1}{m}\frac{\Delta_{1k}}{\Delta} = \frac{Z_{k1}}{m} \qquad [8.27]$$

but all other network quantities are unchanged, i.e.

$$Z'_{kj} = \frac{\Delta'_{jk}}{\Delta'} = \frac{\Delta_{jk}}{\Delta} = Z_{kj}$$

for

$$j, k \neq 1 \qquad [8.28]$$

Now let us see how the elements have been changed by the scaling operation. From [8.22] we see that all mutual admittances of the form y_{1k} ($k \neq 1$) have been multiplied by the factor m. We must now recall from [7.51] that the admittance to ground y_{10} is given by summing the first row or column of the matrix, so that

$$y'_{10} = m^2 y_{11} - m(y_{12} + y_{13} + \ldots + y_{1n}) \qquad [8.29]$$

where, in accordance with [7.51]

$$y_{11} = y_{10} + y_{12} \ldots + y_{1n}$$

Similarly all other admittances to ground are changed, e.g.

$$y'_{20} = y_{22} - m y_{21} - (y_{23} + y_{24} + \ldots + y_{2n}) \qquad [8.30]$$

The network of Fig. 7.11 has now been transformed to that shown in Fig. 8.7.

The question which must be asked at this point is 'under what conditions is the scaled network physically realisable?' Obviously multiplication of an admittance by a passive constant m in no way affects its physical realisability, so that the mutual admittances are all realisable. The only problem concerns the admittances to ground such as [8.29] or [8.30] which involve the differences of two quantities, and may therefore become unrealisable in a passive network formulation. For example, if the y_{jk} are all capacitors and take the form $C_{jk}s$, [8.29] becomes

$$C_{10} = m^2 C_{11} - m(C_{12} + C_{13} + \ldots + C_{1n})$$

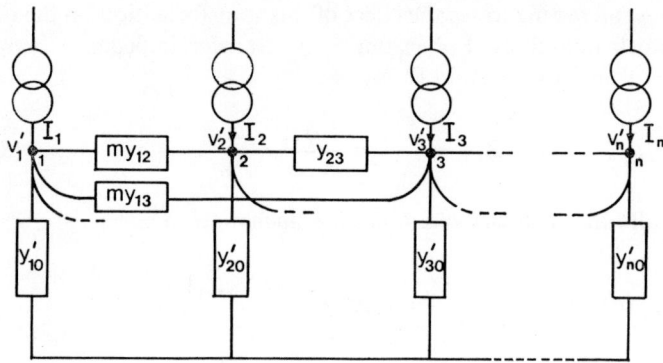

Fig. 8.7. *The effect of scaling port 1 of Fig. 7.11 by the factor m.*

and the limitation on physical realisability is obviously $C_{10} \geqq 0$, i.e.

$$m \geqq \frac{C_{12}+C_{13}+\ldots+C_{1n}}{C_{11}} \qquad [8.31]$$

This has the effect of limiting the amount of impedance transformation which is possible at a given port. However further impedance transformation may often be carried out by scaling other rows and columns of the admittance matrix, which may then result in more suitable values of the circuit elements for a larger impedance transformation at port 1.

8.3.2. IMPEDANCE TRANSFORMATIONS IN TWO-PORT NETWORKS

A useful special case which often arises is that of a two-port network, for example most filter networks are two-ports. Thus we may take Fig. 7.11 or Fig. 8.7 as representing a two-port if we regard node pairs 1, 0 and n, 0 as the two ports. Then we see that impedance-transformations within two-port networks may be carried out by writing down the nodal admittance matrix, and scaling either the first or last row and column pairs. Additional scaling of internal rows and columns may also be carried out. The latter transformations will not affect the port behaviour of the network, but may be useful in allowing more manageable circuit elements to be used. In general an infinite number of *equivalent circuits* may be derived in this fashion.

Example 1. Fig. 8.8 shows a two-port network operating between unity admittances. We wish to find a network with the same driving

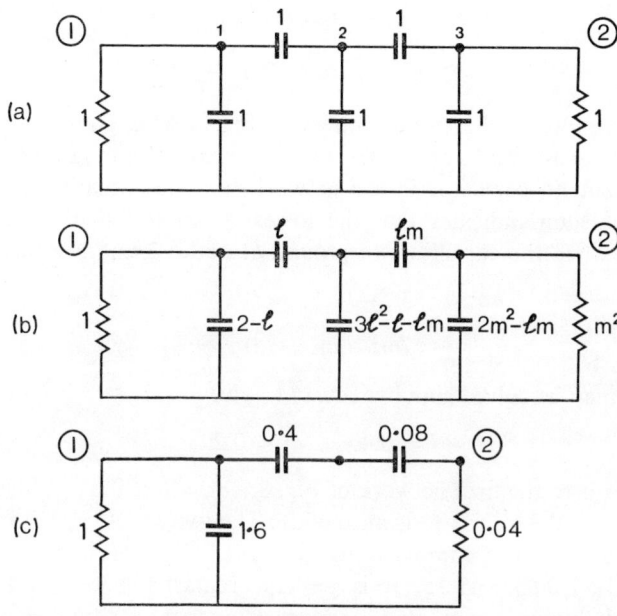

FIG. 8.8. (a) *Network whose nodal admittance matrix is given in [8.32]. (b) The network after scaling; (c) The scaled network with $l = 0.4$, $m = 0.2$.*

point impedance at port 1 but having a minimum value of conductance at port 2. The admittance matrix of the network is

$$\begin{bmatrix} 1+2s & -s & 0 \\ -s & 3s & -s \\ 0 & -s & 1+2s \end{bmatrix} \begin{matrix} l \leftarrow \\ m \leftarrow \\ \end{matrix} \qquad [8.32]$$

$$ l\uparrow \quad \uparrow m$$

In [8.32] we have indicated that the admittance level at node 2 may be raised by multiplying the second row and column by l, and the admittance level at node 3, containing the output conductance, may be raised by the factor m. The scaled admittance matrix becomes

$$\begin{bmatrix} 1+2s & -ls & 0 \\ -ls & 3l^2s & -lms \\ 0 & -lms & 2m^2s+m^2 \end{bmatrix}$$

which results in the circuit shown in Fig. 8.8(b). The problem is to make m as small as possible, and this means choosing both l and m so that no capacitor in the network becomes negative. A little thought soon indicates that the lowest possible value of $m(\neq 0)$ occurs when the capacitors to ground at nodes 2 and 3 vanish, i.e.

$$3l^2 - l - lm = 0$$
$$2m^2 - lm = 0$$

which may be solved simultaneously to give

$$l = 0.4, \quad m = 0.2$$

This leads to the final network of Fig. 8.8(c), where the conductance at port 2 is 0·04, i.e. a resistance of 25. The two series capacitors 0·4 and 0·08 may be combined to give a single capacitor of value $1/(1/0·4+1/0·08) = 1/15$. It is perhaps remarkable that now the network contains only two capacitors compared with the five of the original network.

In order to check the validity of the method it is instructive to work out the driving point impedances at port 1 of the two networks of Fig. 8.8(a) and (c). The reader should be able to verify that in each case we have

$$Z_{11}(s) = \frac{5s+3}{s(8s+5)} \qquad [8.33]$$

Example 2. In the previous example we reduced the number of circuit elements after applying the impedance transformation. In this next example the reverse will occur, but it is chosen to illustrate a very useful technique of impedance transformation. In this example we commence from a circuit consisting simply of a reactive series impedance Z in cascade with an ideal transformer having turns ratio $1:m$, as shown in Fig. 8.9(a). We terminate the circuits in unit

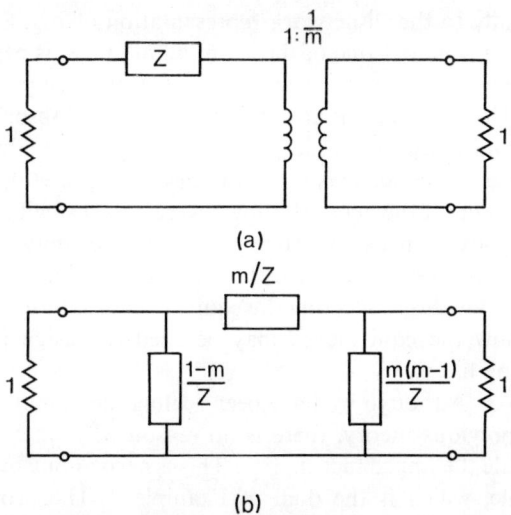

FIG. 8.9. (a) *Original network;* (b) *Equivalent Pi-network obtained through admittance scaling.*

resistors. The object is to eliminate the ideal transformer, but retain a reactive network terminated in unit resistors.

The first step is to absorb the transformer in the load resistor, which becomes a resistance of m^2. Now the admittance matrix may be written down as

$$\begin{bmatrix} 1+\dfrac{1}{Z} & -\dfrac{1}{Z} \\ -\dfrac{1}{Z} & \dfrac{1}{Z}+\dfrac{1}{m^2} \end{bmatrix} \begin{matrix} \\ m \\ \leftarrow \end{matrix} \qquad [8.34]$$

$\uparrow m$

The load conductance at port 2 may now be made unity once more by multiplying the second row and column by m, giving the admittance matrix

$$\begin{bmatrix} 1+\dfrac{1}{Z} & -\dfrac{m}{Z} \\ -\dfrac{m}{Z} & \dfrac{m^2}{Z}+1 \end{bmatrix} \qquad [8.35]$$

leading finally to the Pi-network representation of Fig. 8.9(b). Here it is seen that if $m \neq 1$ one of the shunt admittances is negative with respect to the other, so that one or the other must be physically unrealisable. However there is always the possibility of absorbing such a negative admittance into an adjacent similar positive admittance of greater algebraic value. This is not the case in Fig. 8.9(b), but such a case might occur if the network shown were only a small portion of a more complicated network. Here note that the unity terminating resistors are not essential, since they may be eliminated from both 8.9(a) and (b) without affecting the equivalence established. We shall see later how the equivalence may be used to design impedance-transforming filters.

Example 3. Although we have been scaling the admittance matrix in all the previous theory, there is no reason why we cannot alternatively scale the impedance matrix. This is carried out in the folowing example, which is the dual of Example 2. The problem is to eliminate the transformer of Fig. 8.10(a). Proceeding as in Example 1 we absorb the transformer into the termination at port 2 to give a

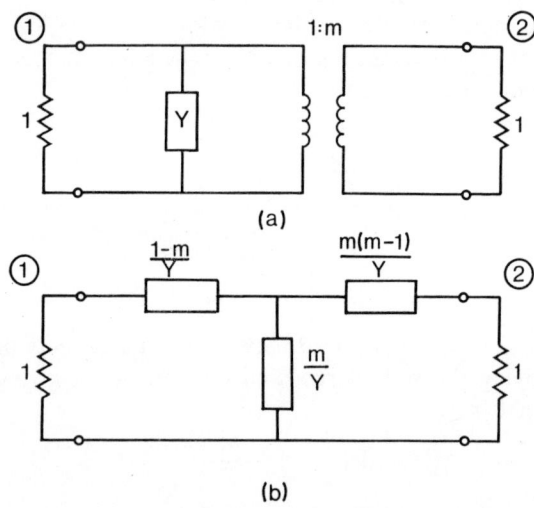

FIG. 8.10. (a) *Original network;* (b) *Equivalent T-network obtained through impedance scaling.*

resistance $1/m^2$, as in Fig. 8.11. This figure indicates how the impedance matrix is formed as

$$\begin{bmatrix} 1+\dfrac{1}{Y} & -\dfrac{1}{Y} \\ -\dfrac{1}{Y} & \dfrac{1}{Y}+\dfrac{1}{m^2} \end{bmatrix} \begin{matrix} \\ m \\ \leftarrow \end{matrix}$$ [8.36]

$$\uparrow m$$

Multiplication of the second row and column by m scales the load resistance at port 2 to unity and gives the impedance matrix

$$\begin{bmatrix} 1+\dfrac{1}{Y} & -\dfrac{m}{Y} \\ -\dfrac{m}{Y} & \dfrac{m^2}{Y}+1 \end{bmatrix}$$ [8.37]

The T-network equivalent of [8.37] is shown in Fig. 8.10(b), and is seen to be the dual of Fig. 8.9(b).

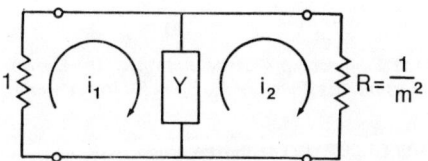

FIG. 8.11. *Pertinent to the formulation of the impedance matrix given in [8.36].*

8.3.3. IMPEDANCE-TRANSFORMING FILTERS

Applications of several of the concepts introduced in this chapter may be illustrated in the next example, which commences from the low-pass prototype filter shown in Fig. 8.12(a). This was introduced also in Section 5.13.3, where the insertion loss was shown to be given by

$$\frac{P_0}{P_L} = 1+\omega^6$$

or

$$\frac{P_0}{P_L} = 10 \log (1+\omega^6) \text{ dB} \qquad [8.38]$$

as illustrated in Fig. 8.12(b).

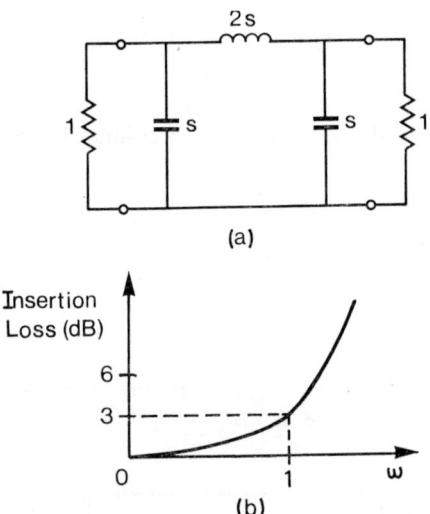

Fig. 8.12. (a) *Low-pass prototype network;* (b) *Corresponding insertion loss given by* $P_0/P_L = 10 \log (1+\omega^6)$.

The problem is to convert this low pass prototype into a band pass filter without transformers, having angular cut-off frequencies of 1000 rad/sec and 3000 rad/sec with terminating resistors of 100 Ω and 625 Ω.

The first thing to realize is that it is impossible to introduce the ideal transformer into the low pass prototype and then eliminate it by admittance matrix scaling. This can be seen by applying the method directly, or by inspection from the network equivalences shown in Figs. 8.9 and 8.10. For example if we follow the second shunt capacitor in Fig. 8.12(a) with a 1:m ideal transformer and apply the equivalence of Fig. 8.10, positive and negative series capacitors will result. There is no adjacent positive capacitor in series to absorb the negative capacitor, and so the method does not give a realizable network. This must be so from physical considerations, because the low-pass prototype is perfectly matched between equal resistances at zero frequency, and it is impossible to match two *unequal* resistances at zero frequency. Transformers do not operate at D.C. Therefore the first step must be to carry out the low-pass to band-pass transformation given in Section 8.2.4 as [*8.17*] using [*8.15*] and [*8.18*]. This will result in a network having a multiple pole of attenuation at D.C., and eliminates one obstacle to introducing impedance transformation.

In order to keep the numbers occurring in the calculations close to unity it is convenient to carry out the low-pass to band-pass transformation in two steps. The first step is to transform the low pass filter to give $\omega_1 = 1/\sqrt{3}$, and $\omega_2 = \sqrt{3}$, so that $\omega_2/\omega_1 = 3$ and $\omega_0 = \sqrt{(\omega_1\omega_2)} = 1$. This gives the correct bandwidth for the band pass filter. The second step will be to scale the frequency by a factor of $1000\sqrt{3}$. In applying the first step [*8.17*] and [*8.18*] become

$$s \to Q[s+(1/s)]$$

with

$$Q = \frac{\omega_0}{\omega_2-\omega_1} = \frac{1}{\sqrt{3}-1/\sqrt{3}} = \frac{\sqrt{3}}{2}$$

giving

$$s \to (\sqrt{3}/2)[s+(1/s)] \qquad [8.39]$$

Applying the transformation of [*8.39*] to the low pass prototype of Fig. 8.12(*a*) gives the band pass filter of Fig. 8.13(*a*) having the response indicated in Fig. 8.13(*b*).

There are several ways of introducing the 6·25:1 impedance transformation into the network. The method used here is to make use of the equivalence shown in Fig. 8.10(*a*). Hence we wish to introduce an ideal transformer of turns ratio 1:m adjacent to a shunt element, and we chose the right-hand shunt inductor of admittance $\sqrt{3}/2s$ of Fig. 8.13(*a*). This must be immediately cancelled by a transformer of equal and opposite turns ratio, as indicated in Fig. 8.14(*a*). We can now absorb the right-hand transformer into the network, which raises the impedance level of the network to the right of the transformer by a factor m^2, as shown in Fig. 8.14(*b*). The first transformer is combined with the shunt admittance $\sqrt{3}/2s$ to give the T-network equivalent shown in Fig. 8.14(*b*). We see that if $m > 1$ the

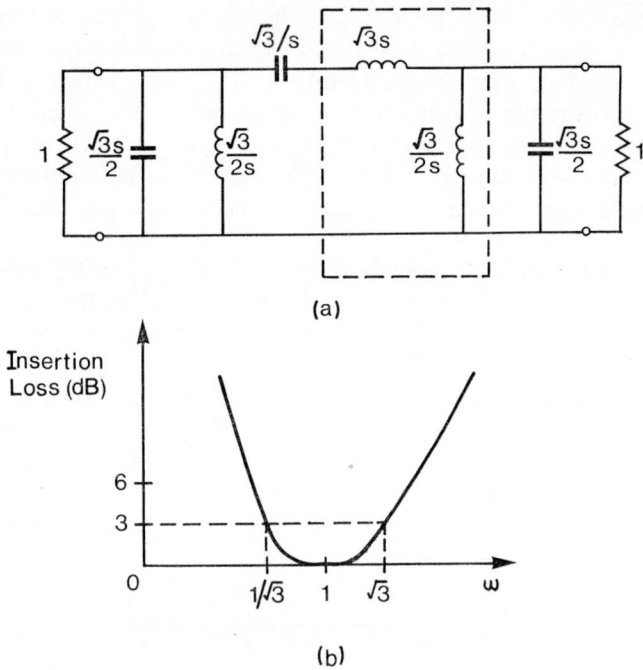

FIG. 8.13. (a) *Band-pass filter prototype;* (b) *Corresponding insertion loss.*

first element of the T is negative and is to be absorbed into the adjacent series element. This is possible only if

$$\sqrt{3}s + \frac{2(1-m)}{\sqrt{3}} s \geqq 0$$

i.e. if $m \leqq 2\cdot 5$.

We require a turns ratio of 2·5:1 to give a 6·25:1 impedance transformation, so the original question allowed the solution to be obtained using just the one circuit transformation. Substituting $m = 2\cdot 5$ gives the transformed network of Fig. 8.14(c). It is possible to obtain further impedance transformation, but it is not required in this example.

Finally we are required to denormalise the network of Fig. 8.14(c) to give band edge frequencies at 1000 rad/sec and 3000 rad/sec and terminating impedances of 100Ω and 625Ω. This requires frequency

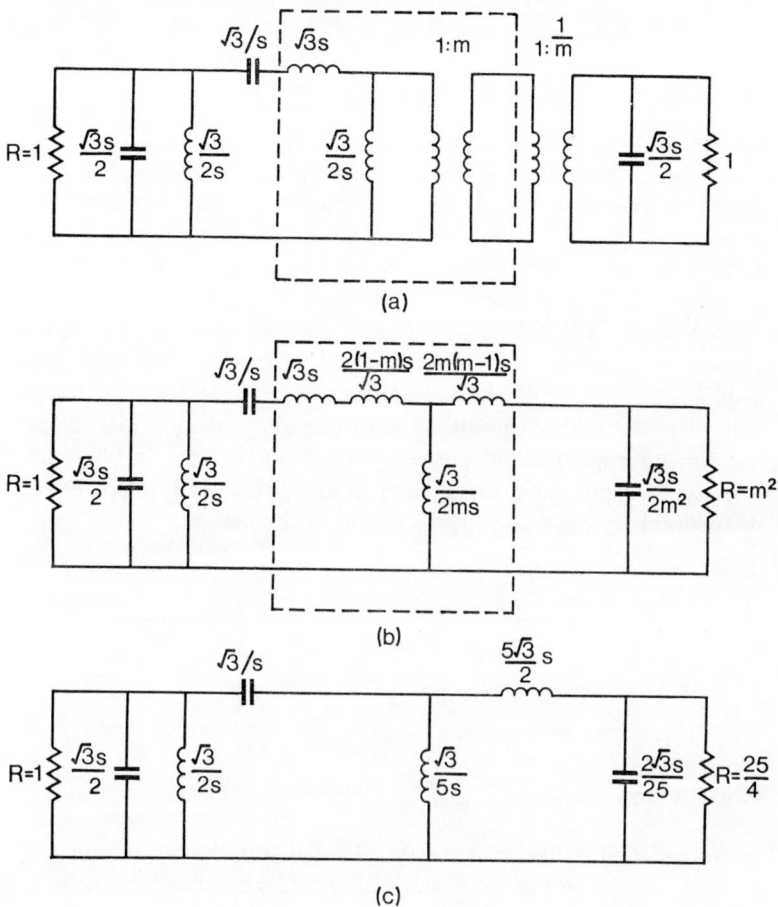

Fig. 8.14. *Successive stages in the introduction of impedance transformation into the filter of Fig. 8.13(a).*

to be scaled by a factor $\omega_0 = 1000\sqrt{3}$ and impedance by a factor $R = 100$. As stated in Section 8.1, all resistances are scaled by R, all inductances are multiplied by $R/\omega_0 = 1/10\sqrt{3}$, and all capacitances are divided by $R\omega_0 = \sqrt{3} \times 10^5$. The final values of the circuit elements are of no interest here, and the final circuit element values will not be calculated specifically.

8.4. Transfer Phase and Group Delay

8.4.1. TRANSFER PHASE OF A TWO-PORT

In the previous discussion of filters we have been concerned with tailoring their design to give a desired amplitude response. It is now time to turn our attention to the phase response of the filter networks, which is frequenctly of concern in filter design. The object of a filter is to transmit signals in a narrow frequency band as faithfully as possible, while suppressing other frequencies. However the transmitted signal will suffer distortion if the phase versus frequency response of the network is nonlinear. Often this is of little consequence, e.g. in radar receivers where one is concerned with reception of pulses without primarily worrying about their actual shape, but for other applications phase is very important, e.g. in filters for television signals (and particularly colour television) where phase distortion may cause serious distortion to the picture.

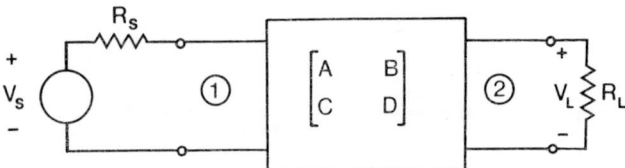

FIG. 8.15. *Doubly-terminated two-port*.

We will first define the transfer phase of the doubly terminated two-port shown in Fig. 8.15. In Section 5.13 we defined the voltage transfer function or voltage insertion ratio as the ratio of the voltage V'_L across the load resistor where the network is replaced by an ideal transformer giving maximum power transfer, to the actual voltage V_L across the load resistor when the network is present. In terms of the transfer matrix parameters of the network the final equation was given in [5.112], i.e.

$$\frac{V'_L}{V_L} = \tfrac{1}{2}\left(A\sqrt{\left(\frac{R_L}{R_s}\right)} + D\sqrt{\left(\frac{R_s}{R_L}\right)} + \frac{B}{\sqrt{(R_L R_s)}} + C\sqrt{(R_L R_s)}\right)$$

[8.40]

CIRCUIT TRANSFORMATION AND FILTERS

and we defined the insertion loss of the two-port as

$$\frac{P_0}{P_L} = \left|\frac{V'_L}{V_L}\right|^2 \qquad [8.41]$$

In the case when the source and load resistors are matched by an ideal transformer giving the voltage V'_L the phase of the signal and load voltages are the same, since there is no phase shift across an ideal transformer. Most other two port networks will introduce some phase shift however. The phase is defined at real frequencies when $s = j\omega$, when [8.40] takes the form

$$V'_L/V_L = R + j\mathscr{I} \qquad [8.42]$$

where R and $j\mathscr{I}$ are the real and imaginary terms respectively. The *transfer* or *insertion phase* of the two port between the source and load resistors is now defined as

$$\psi = \tan^{-1}(\mathscr{I}/R) \qquad [8.43]$$

which is simply the phase of the complex number given by [8.42]. Note that if either R or $\mathscr{I}$ is zero for all values of ω, then ψ is a constant. A typical such case is an ideal transformer where

$$A = 1/n, \qquad D = n, \qquad B = C = 0,$$

giving

$$\frac{V'_L}{V_L} = \tfrac{1}{2}\left(\frac{1}{n}\sqrt{\left(\frac{R_L}{R_s}\right)} + n\sqrt{\left(\frac{R_s}{R_L}\right)}\right) \qquad [8.44]$$

which is real and frequency independent.

A more interesting general case is the three element filter which have been using frequently for the demonstration of filter properties, e.g. in Figs. 5.23 and 8.12. The transfer matrix of the three reactive circuit elements is given by multiplying the individual transfer matrices, i.e.

$$\begin{bmatrix} 1 & 0 \\ s & 1 \end{bmatrix} \begin{bmatrix} 1 & 2s \\ 0 & 1 \end{bmatrix} \begin{bmatrix} 1 & 0 \\ s & 1 \end{bmatrix} = \begin{bmatrix} 2s^2+1 & 2s \\ 2s^3+2s & 2s^2+1 \end{bmatrix}$$

The voltage transfer function between the unity source and load resistances is given from [8.40] as

$$V'_L/V_L = (2s^2+1) + (s^3+2s)$$

The right hand side has been separated into even and odd polynomial terms because when we consider the real frequency behaviour of the transfer phase (by substituting $s = j\omega$) the even polynomial is real and the odd polynomial becomes the imaginary part. In fact [8.43] gives

$$\psi = \tan^{-1} \frac{s^3+2s}{j(2s^2+1)}\bigg|_{s=j\omega}$$

$$= \tan^{-1} \frac{\omega^3-2\omega}{2\omega^2-1} \qquad [8.45]$$

This may be generalised to the case of any lossless two port, where the A and D terms of the transfer matrix are always real, and the B and C terms always imaginary at real frequencies. This statement implies that, A and D are even, B and C are odd polynomials, and

$$\psi \text{ (lossless)} = \tan^{-1} \frac{B+C}{j(A+D)}\bigg|_{s=j\omega} \qquad [8.46]$$

8.4.2. GROUP VELOCITY AND GROUP DELAY

The phase response represented by [8.45] is rather complicated, and it would be desirable to express its effect on a transmitted signal in a rather simpler way. This is accomplished by the introduction of the concept of time delay or group delay, which is commonly introduced in physical wave theory, in consequence of which the reader is possibly familiar with the topic from studies in physics. In the latter the concept is introduced initially by consideration of a sinusoidal wave propagating in a medium represented by a function of space z and time t, i.e.

$$\psi(z, t) = A \cos(\beta z - \omega t) \qquad [8.47]$$

Here $\psi(z, t)$ is the wave function, A is the wave amplitude, and $(\beta z - \omega t)$ is the phase. Surfaces of constant phase are defined by

$$\beta z - \omega t = \text{constant} \qquad [8.48]$$

and hence such surfaces propagate with the velocity

$$v = \frac{dz}{dt} = \frac{\omega}{\beta} \qquad [8.49]$$

However, purely sinusoidal waves carry no information, which must be impressed by means of amplitude- or phase-modulation techniques. A modulated wave cannot be represented in the simple sinusoidal form [*8.45*] and the phase velocity tends to lose its precise significance. We can see this by considering the superposition of two sinusoidal waves which differ slightly in frequency ω and propagation function β, i.e.

$$\psi_1 = \cos(\beta z - \omega t)$$
$$\psi_2 = \cos[(\beta + \Delta\beta)z - (\omega + \Delta\omega)t]$$

whose sum is

$$\psi = \psi_1 + \psi_2 = 2\cos\tfrac{1}{2}(\Delta\beta z - \Delta\omega t)\cos[(\beta + \tfrac{1}{2}\Delta\beta)z - (\omega + \tfrac{1}{2}\Delta\omega)t] \qquad [8.50]$$

This composite waveform represents a wave having slightly different β and ω than ψ_1, but the amplitude is given by

$$A = 2\cos\tfrac{1}{2}(\Delta\beta z - \Delta\omega t) \qquad [8.51]$$

which varies slowly in both space and time between values of 0 and ± 2, as indicated in Fig. 8.16. This waveform may now be con-

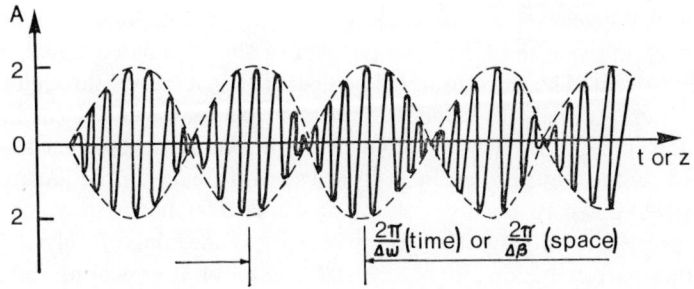

FIG. 8.16. *Pertinent to the physical definition of group velocity.*

sidered to carry information, in this case in the form of periodic beats, groups, or wave packets, as they are variously known. The surfaces of constant phase for these groups are defined by the equation

$$\Delta\beta z - \Delta\omega t = \text{constant}$$

and hence they propagate with the velocity

$$v_g = \frac{dz}{dt} = \frac{\Delta\omega}{\Delta\beta} \qquad [8.52]$$

In the case when $\Delta\omega$ and $\Delta\beta$ are very small, we may write [8.52] in terms of differential coefficients to give

$$v_g = d\omega/d\beta \qquad [8.52]$$

The reciprocal of the group velocity has the dimensions of time, and is known as the group or time delay, i.e.

$$\tau = d\beta/d\omega \qquad [8.53]$$

and is the time delay in a unit length of the medium. In a case where β and ω are linearly related by a relationship such as

$$\beta = K\omega \qquad [8.54]$$

then the phase velocity given by [8.49] and the group velocity given by [8.52] are equal. This is the definition of a dispersionless medium. In such a medium a wave packet will propagate without distortion. However in a medium where β and ω are not linearly related the group velocity is a function of ω. We know that a modulated waveform necessarily possesses a finite bandwidth, and the greater the information to be transmitted the greater must be the bandwidth. Hence the velocity or time delay of the various component frequencies comprising the Fourier transform of the modulated signal will be different. The waveform will be distorted as it travels through the medium. Actually the velocity of the wave packet may not be defined in the case of a rather wide band signal in a very dispersive medium, because the wave packet loses its shape as it travels in the medium, and there can be no well defined or stable reference with which to measure the velocity. The group velocity is meaningful only for a rather narrow bandwidth wherein $d\beta/d\omega$ does not vary appreciably. Alternatively if $d\beta/d\omega$ is fairly constant over a broad bandwidth then we will be able to apply broad band modulation with full confidence that the signal will not suffer appreciable distortion.

8.4.3. GROUP DELAY OF A TWO-PORT NETWORK

Everything we defined in the previous section is fully applicable to an electric wave travelling in a circuit, which is the 'medium'. For

CIRCUIT TRANSFORMATION AND FILTERS

example, returning to consideration of the phase properties of the two port of Fig. 8.12 as given by [8.45], the group delay is given by application of [8.53] as

$$\tau_a = \frac{d\psi}{d\omega}$$

$$= \frac{1}{1+\left(\dfrac{\omega^3-2\omega}{2\omega^2-1}\right)^2} \cdot \frac{(2\omega^2-1)(3\omega^2-2)-(\omega^3-2\omega)4\omega}{(2\omega^2-1)^2}$$

$$= \frac{2\omega^4+\omega^2+2}{\omega^6+1} \qquad [8.55]$$

[8.53] represented the time delay of the group in a unit length of the medium, but [8.55] is the total time delay from port 1 to port 2 of the circuit since the phase ψ is the total phase shift across the network, and is not necessarily referenced to a physical length. Alternatively if the circuit had physical length, as it could have for example if the circuit included transmission lines, and if we could define a propagation coefficient β and a physical length l, then the ψ in [8.55] would be given by $\psi = \beta l$, so that the group delay in [8.55] would indeed be the actual time delay of the group through the network.

We see that [8.55] is the ratio of two rational polynomials and is a much simpler expression than the phase, [8.45]. ψ and τ are plotted as a function of ω in Fig. 8.17, and it is obvious that the delay plot is a much more sensitive indication of the dispersion properties of the network than the phase plot. The latter is almost linear, and it is difficult to see the small changes in the slope of the curve. These small changes are not really too small in terms of group delay, which is seen to vary appreciably over the low-pass band of the filter. Actually this three-element filter is not really a very selective filter, and in the case of practical highly selective filters the group delay varies typically by a factor of 2 or more over the pass band.

8.5. All-pass Networks

In the previous section we learnt that if we design a frequency-selective filter to give a good amplitude response over a band of frequencies, then the group delay may vary appreciably over that band and give rise to signal distortion, even though the amplitude

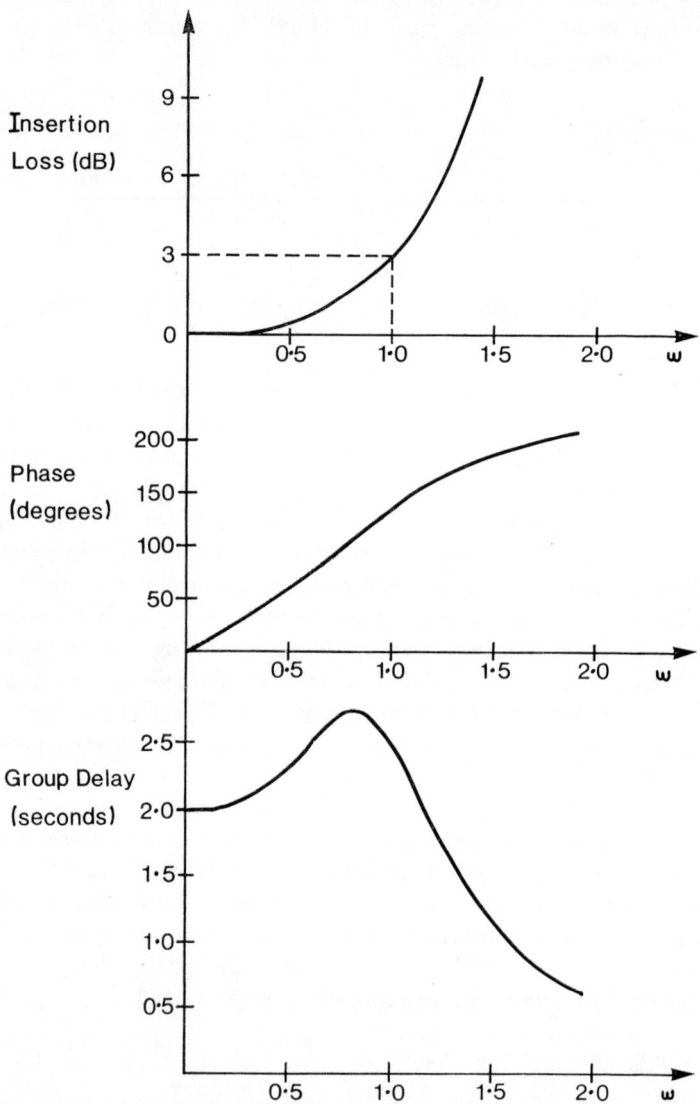

FIG. 8.17. *Characteristics of a low-pass filter* (a) *Insertion loss* = $10 \log(1+\omega^6)$; (b) *Phase angle*, ψ, *as given in* [8.45]; (c) *Group delay* τ as given in [8.55].

CIRCUIT TRANSFORMATION AND FILTERS 355

response is 'flat'. This situation may be corrected either by designing the filter to give a linear phase response, and hence a constant group delay, or by following the filter with a network which compensates the group delay characteristic while not affecting the amplitude. The first technique is an advanced topic which we shall discuss again in Chapter 12. In fact the most commonly employed method is to use the external phase equalisation network, which because (ideally) it does not introduce further amplitude variation is known also as an all-pass network.

8.5.1. SYMMETRICAL LATTICE NETWORKS

A basic circuit for the derivation of all-pass networks is shown in Fig. 8.18(a) as a bridge network containing two impedances Z_a and two impedances Z_b. Terminals 1, 1' and 2, 2' are selected to form a two-port network. This is now re-drawn in the form shown in Fig. 8.18(b) with the ports on the left and right hand sides, which is a more convenient format when we begin to cascade such networks. Only two of the impedances are shown, the presence of the other two being indicated by dotted lines. Thus we have an impedance Z_a connected between terminals 1' and 2' and an impedance Z_b connected between terminals 1' and 2. This notation for a *symmetrical lattice network* is widely employed throughout the literature.

In order to establish the properties of the symmetrical lattice it is necessary first to find the two-port parameters of this network. The open-circuit impedance parameters defined in Section 5.2 may be derived quite readily. Thus we have

$$z_{11} = \frac{V_1}{I_1}\bigg|_{I_2=0} = \tfrac{1}{2}(Z_a+Z_b) \qquad [8.56]$$

for with an open circuit at port 2 the input impedance consists of two impedances Z_a+Z_b in parallel. Since the network is symmetrical, $z_{22} = z_{11}$, and it remains to find

$$z_{12} = z_{21} = \frac{V_2}{I_2}\bigg|_{I_2=0} \qquad [8.57]$$

This is accomplished easily if we note that the potential at terminals 2 and 2' may be found by using the principle of the potential divider with the input voltage V_1 being applied across Z_a+Z_b in the two distinct paths 1, 2, 1' and 1, 2', 1'. Thus the potential at terminal 2

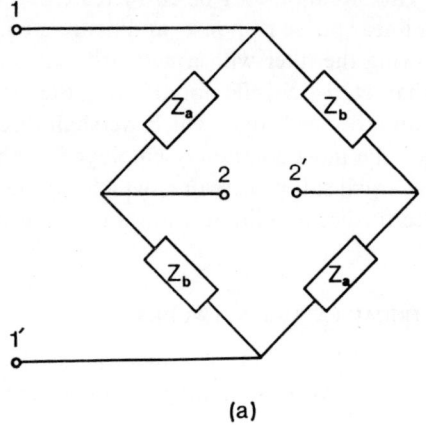

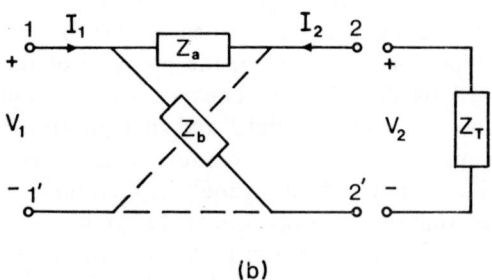

FIG. 8.18. *Symmetrical lattice network* (a) *in bridge form;* (b) *in lattice form.*

relative to that at 1' is $Z_b V_1/(Z_a+Z_b)$, and the potential at terminal 2' relative to 1' is $Z_a V_1/(Z_a+Z_b)$. Hence the potential difference between terminals 2 and 2' is

$$V_2 = \frac{Z_b - Z_a}{Z_b + Z_a} V_1 \qquad [8.58]$$

This relationship enables us to form

$$z_{12} = z_{21} = \frac{V_2}{I_1}\bigg|_{I_2=0} = \frac{V_2}{V_1} \cdot \frac{V_1}{I_1}\bigg|_{I_2=0}$$
$$= \tfrac{1}{2}(Z_b - Z_a) \qquad [8.59]$$

The open-circuit impedance matrix is therefore

$$(z) = \tfrac{1}{2}\begin{bmatrix} Z_b+Z_a & Z_b-Z_a \\ Z_b-Z_a & Z_b+Z_a \end{bmatrix} \quad [8.60]$$

The symmetrical lattice has an interesting additional symmetry property which appears if we also derive the short circuit admittance matrix, as defined in Section 5.5. First substitute

$$Y_a = 1/Z_a, \quad Y_b = 1/Z_b \quad [8.61]$$

and short circuit terminals 2 and 2'. The circuit is re-drawn in Fig. 8.19 to show the input current I_2 splitting into currents i_1 through Y_a and i_2 through Y_b. We see that

$$i_1 = \frac{Y_a}{Y_a+Y_b} I_1; \quad i_2 = \frac{Y_b}{Y_a+Y_b} I_1$$

giving

$$I_2 = i_2 - i_1 = \frac{Y_b - Y_a}{Y_b + Y_a} I_1 \quad [8.62]$$

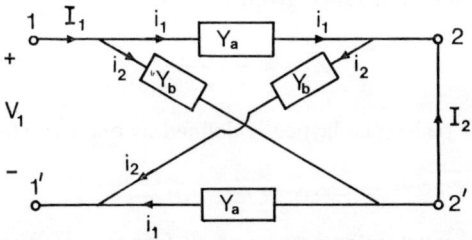

FIG. 8.19. *Pertinent to the derivation of the two-port admittance matrix of a symmetrical lattice network.*

The short circuit driving point admittance is

$$y_{11} = \frac{I_1}{V_1}\bigg|_{V_2=0} = \tfrac{1}{2}(Y_b + Y_a) \quad [8.63]$$

and the short circuit transfer admittance is

$$y_{12} = y_{21} = \frac{I_2}{V_1}\bigg|_{V_2=0} = \frac{I_2}{I_1} \cdot \frac{I_1}{V_1}\bigg|_{V_2=0}$$

which by application of [8.62] and [8.63] gives

$$y_{12} = y_{21} = \tfrac{1}{2}(Y_b - Y_a) \qquad [8.64]$$

Hence the short circuit admittance matrix of the symmetrical lattice network is

$$(y) = \tfrac{1}{2}\begin{bmatrix} Y_b + Y_a & Y_b - Y_a \\ Y_b - Y_a & Y_b + Y_a \end{bmatrix} \qquad [8.65]$$

which is the same as [8.60] in appearance. It is interesting to check that (z) and (y) are related through $(y) = (z)^{-1}$, as in [5.38].

8.5.2. THE CONSTANT RESISTANCE SYMMETRICAL LATTICE

The next step is to terminate port 2 with an impedance Z_T as indicated in Fig. 8.18(b). Application of [5.42] gives the driving point impedance

$$Z_{11} = \frac{\det z + z_{11} Z_T}{Z_T + z_{22}} \qquad [8.66]$$

and the transfer impedance given by [5.43] is

$$Z_{21} = \frac{z_{21} Z_T}{Z_T + z_{22}} \qquad [8.67]$$

A constant resistance lattice is defined as one for which

$$Z_{11} = Z_T = R \qquad [8.68]$$

where R is a constant resistance at all frequencies. From [8.60] we see that

$$\det z = \tfrac{1}{4}[(Z_b + Z_a)^2 - (Z_b - Z_a)^2]$$
$$= Z_a Z_b \qquad [8.69]$$

and substituting [8.68] and [8.69] into [8.66] gives

$$Z_{11} = \frac{Z_a Z_b + \tfrac{1}{2}(Z_b + Z_a)R}{R + \tfrac{1}{2}(Z_b + Z_a)} = R$$

leading to the important condition

$$Z_a Z_b = R^2 \qquad [8.70]$$

CIRCUIT TRANSFORMATION AND FILTERS

This equation states that in order for the input impedance to equal the constant terminating resistance R at all frequencies, it is necessary only for Z_a and Z_b to be reciprocally related, i.e. that

$$Z_b = R^2/Z_a \qquad [8.71]$$

For example if Z_a is a capacitance, Z_b will be an inductance.

Turning now to consideration of the transfer impedance, [8.67] becomes

$$Z_{12} = Z_{21} = \frac{\frac{1}{2}(Z_b - Z_a)R}{R + \frac{1}{2}(Z_b + Z_a)}$$

and substituting for Z_b using [8.71] we have

$$Z_{21} = \frac{R^2/Z_a - Z_a}{R^2/Z_a + 2R + Z_a} = \frac{R^2 - Z_a^2}{(R + Z_a)^2}$$

$$= \frac{R - Z_a}{R + Z_a} \qquad [8.72]$$

Since the constant resistance lattice network is perfectly matched, it is possible to cascade several such lattices each with the same resistance R and the cascade will still be matched. Hence the network will have perfect transmission as far as amplitude is concerned, and only the phase of the transmitted signal will change. This is a basis then for the design of all-pass networks.

In order to represent the transmission properties of a cascade of constant resistance lattices we shall first form the transfer matrix of the network. The transfer matrix is found by application of [5.55] to give

$$(T) = \begin{bmatrix} \dfrac{Z_b + Z_a}{Z_b - Z_a} & \dfrac{2Z_a Z_b}{Z_b - Z_a} \\ \dfrac{2}{Z_b - Z_a} & \dfrac{Z_b + Z_a}{Z_b - Z_a} \end{bmatrix} \qquad [8.73]$$

and in terms of image parameters as defined in the case of a symmetrical network by [5.14] we see that

$$Z_I = \sqrt{(B/C)} = \sqrt{(Z_a Z_b)} = R \qquad [8.74]$$

where we have used [*8.70*] and find an image impedance equal to R, as expected. [*5.141*] gives the propagation function γ as

$$\begin{aligned}
\exp(\gamma) &= \cosh\gamma + \sinh\gamma = A + \sqrt{(BC)} \\
&= (Z_b + Z_a + 2\sqrt{(Z_a Z_b)})/(Z_b - Z_a) \\
&= (R^2/Z_a + Z_a + 2R)/(R^2/Z_b - Z_a) \\
&= (R+Z_a)^2/(R^2 - Z_a^2) \\
&= 1/Z_{21}
\end{aligned} \qquad [8.75]$$

where in the final line we have made use of the formula for the transfer impedance given by [*8.72*]. Hence the propagation function is given by

$$\gamma = -\ln Z_{12} \qquad [8.76]$$

In Section 5.14.1 it was shown that if we cascade image-matched two ports then the image impedance of the overall two-port is equal to the image impedance of any individual two port (which is equal to R), and the overall propagation function is the sum of the individual

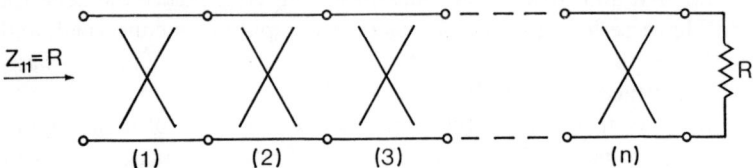

FIG. 8.20. *Cascade of n constant-resistance lattices.*

propagation functions. Hence for the n constant resistance lattices shown in Fig. 8.20 the propagation coefficient of the cascade is

$$\gamma_{1n} = \gamma^{(1)} + \gamma^{(2)} + \ldots + \gamma^{(n)}$$

which from [*8.76*] gives

$$\gamma_{1n} = -\ln[Z_{12}^{(1)} Z_{12}^{(2)} \ldots Z_{12}^{(n)}] \qquad [8.77]$$

If the overall transfer impedance is Z_{12}, we have

$$Z_{12} = Z_{12}^{(1)} Z_{12}^{(2)} \ldots Z_{12}^{(n)} \qquad [8.78]$$

since the overall propagation function $\gamma = -\ln(Z_{12})$ by definition. This result [8.78] states that the overall transfer impedance of a cascade of constant resistance lattices each of resistance R is the product of the transfer impedances of the individual lattices. This result is an obvious simplification in the synthesis of all-pass networks, where a required transfer impedance of high order may be broken down into a product of simple transfer functions, which may then be realised directly and cascaded to form the required network.

One other result to be used later is the voltage transfer function of [5.112] which is derived for the transfer matrix [8.73] as

$$\frac{V'_L}{V_L} = \frac{1}{(Z_b - Z_a)}(Z_b + Z_a + Z_a Z_b/R + R)$$

(since $R_L = R_s = R$), and substituting for Z_b as [8.71] leads to

$$\frac{V'_L}{V_L} = \frac{R^2/Z_a + Z_a + 2R}{R^2/Z_a - Z_a} = \frac{R + Z_a}{R - Z_a} \qquad [8.79]$$

This is seen to be equal to $\exp(\gamma)$ as given by [8.75], and also the reciprocal of the transfer impedance Z_{21}. Hence the phase of the voltage transfer function is the negative of that of the transfer impedance.

8.5.3. THE C-SECTION

The simplest type of reactive constant resistance lattice consists of two inductances and two capacitances, as shown in Fig. 8.21(a). Hence we have

$$Z_a = Ls; \qquad Z_b = R^2/Ls \qquad [8.80]$$

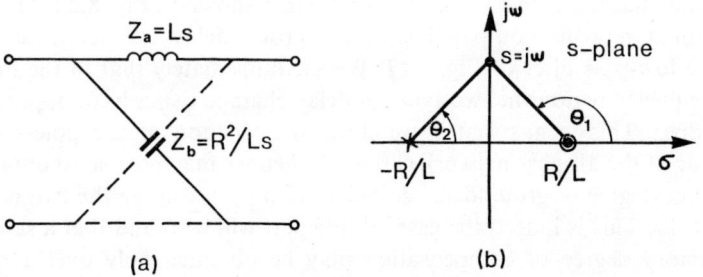

FIG. 8.21. (a) *An all-pass C-section;* (b) *Corresponding pole-zero pattern.*

and the transfer impedance of [8.72] becomes

$$Z_{21} = \frac{R-Ls}{R+Ls} \qquad [8.81]$$

This has the pole-zero pattern shown in Fig. 8.21(b), i.e. a pole at $s = -R/L$ and a zero at $s = +R/L$. The real frequency behaviour of Z_{21} may be found by the 'vector' method described in Section 7.4. The two vectors joining a point on the real frequency axis to the pole and zero are shown. The modulus of Z_{21} is the ratio of these two vector amplitudes, which, since these are equal for all ω, is always equal to unity. This is obvious also from [8.81], and from the fact that we are dealing here with an all-pass network, and have synthesised this properly at the outset. The phase of Z_{21} is equal to

$$\psi = \arg\{Z(j\omega)\} = \theta_1 - \theta_2 = -2\tan^{-1}(\omega L/R) \qquad [8.82]$$

which result is obtained directly either from [8.81] or geometrically from Fig. 8.21(b). The result in the previous section is now applied and gives the phase of the voltage transfer function as $-\psi$. The group delay of this all-pass network as defined in Section 8.4.3 is

$$\tau(\omega) = -\frac{d\psi}{d\omega} = \frac{2L/R}{1+(\omega L/R)^2} \qquad [8.83]$$

At D.C. this has a value $\tau_0 = 2L/R$, and [8.83] may be written as

$$\frac{\tau(\omega)}{\tau_0} = \frac{1}{1+(\omega L/R)^2} \qquad [8.84]$$

which has the type of characteristic shapes shown in Fig. 8.22. These should now be compared with the group delay characteristic of the low-pass filter of Fig. 8.17. We see immediately that in the low frequency region the two types of delay characteristics have opposite slopes. This means that if we chose the location of the pole-zero pair of the all-pass network carefully, then we may be able to obtain some degree of group delay compensation by cascading the two networks. This is indeed the case, although it will be found that a satisfactory degree of compensation may be obtained only over a restricted portion of the low-pass band, typically from $\omega = 0$ to $\omega = 0.6\omega_c$ where ω_c is the cut off frequency.

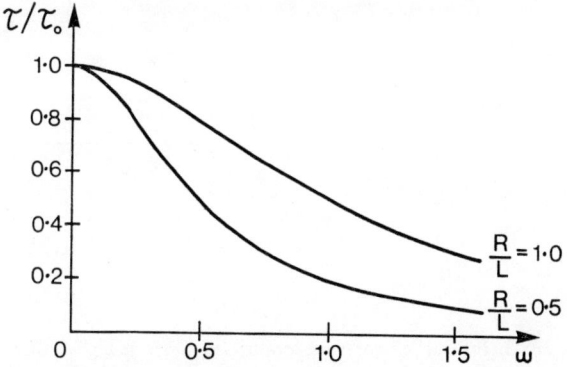

FIG. 8.22. *Normalised group-delay characteristics of all-pass C-sections.*

An all-pass network consisting of a single pole-zero pair symmetrically located on the real axis in the s-plane is known as a C-section. The next highest degree of complication is an all-pass network consisting of a quadruplet of two poles and two zeros, this being known as an all-pass D-section. Any more-complicated type of all-pass function may always be synthesised as a cascade of C- and D-sections.

8.5.4. THE ALL-PASS D-SECTION

Here we take our basic constant-resistance symmetrical lattice of Fig. 8.18 and select

$$Z_a = Ls + 1/Cs \qquad [8.85]$$

so that

$$Y_b = 1/Z_b = Z_a/R^2 \qquad [8.86]$$

giving the lattice of Fig. 8.23(a). The transfer impedance is

$$Z_{12} = \frac{R - Z_a}{R + Z_a} = \frac{R - (Ls + 1/Cs)}{R + (Ls + 1/Cs)}$$

$$= \frac{s^2 - (R/L)s + 1/LC}{s^2 + (R/L)s + 1/LC} \qquad [8.87]$$

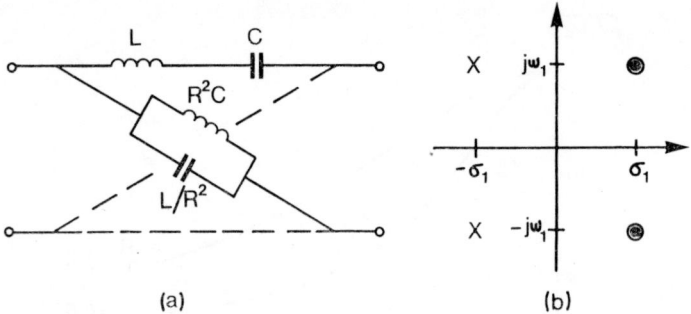

Fig. 8.23. (a) *An all-pass D-section;* (b) *Corresponding pole-zero pattern.*

This is the ratio of two polynomials each of second degree, which we should expect, since each element in the symmetrical lattice is a second-degree network. The poles and zeros are given by

$$s = \pm R/2L \pm \sqrt{[(R/2L)^2 - 1/LC]} \qquad [8.88]$$

As usual with second-degree networks there are two main cases of interest depending on whether $(R/2L)^2$ is greater or less than $1/LC$. In the case where

$$(R/2L)^2 \geqq 1/LC \qquad [8.89]$$

then all four poles and zeros are real. In the case of the equals sign there are two coincident poles and two coincident zeros, all on the real axis. Hence when [8.89] applies the D-section is equivalent to a cascade of two C-sections. Another way to look at this is that [8.87] may then be written as

$$Z_{12} = \frac{(s-\sigma_1)(s-\sigma_2)}{(s+\sigma_1)(s+\sigma_2)} \qquad [8.90]$$

where σ_1, σ_2 are real and positive. This may be broken up into a cascade of two C-sections with

$$Z_{12}^{(1)} = \frac{s-\sigma_1}{s+\sigma_1}; \qquad Z_{12}^{(2)} = \frac{s-\sigma_2}{s+\sigma_2} \qquad [8.91]$$

because of our previous result expressed in [8.78].

CIRCUIT TRANSFORMATION AND FILTERS

The all-pass D-section is of much greater interest when [8.89] does not apply, and we then have a quadruplet of complex poles and zeros, [8.88] becoming

$$s = \pm\sigma_1 \pm j\omega_1$$

with

$$\sigma_1 = R/2L, \quad \omega_1 = \sqrt{(1/LC - (R/2L)^2)}$$

$$(R < 2\sqrt{(L/C)}) \qquad [8.92]$$

The pole-zero pattern is shown in Fig. 8.23(b). The transfer function is

$$Z_{12} = \frac{(s-\sigma_1+j\omega_1)(s-\sigma_1-j\omega_1)}{(s+\sigma_1+j\omega_1)(s+\sigma_1-j\omega_1)}$$

$$= \frac{(s-\sigma_1)^2 + \omega_1^2}{(s+\sigma_1)^2 + \omega_1^2}$$

$$= \frac{s^2 + \sigma_1^2 + \omega_1^2 - 2\sigma_1 s}{s^2 + \sigma_1^2 + \omega_1^2 + 2\sigma_1 s} \qquad [8.93]$$

This has a phase function at $s = j\omega$ given by

$$\psi = -2\tan^{-1}\frac{2\sigma_1\omega}{(\sigma_1^2+\omega_1^2)-\omega^2} \qquad [8.94]$$

The resonant frequency of the series LC and shunt LC networks in the lattice under consideration is given by

$$\omega_0 = 1/\sqrt{(LC)} \qquad [8.95]$$

and from [8.92] we see that

$$\omega_1^2 = \omega_0^2 - \sigma_1^2 \qquad [8.96]$$

This enables us to write [8.94] more simply as

$$\psi = -2\tan^{-1}\frac{2\sigma_1\omega}{\omega_0^2-\omega^2} \qquad [8.97]$$

The group delay is given by

$$\tau = -\frac{d\psi}{d\omega} = \frac{4\sigma_1}{1+\left(\dfrac{2\sigma_1\omega}{\omega_0^2-\omega^2}\right)^2} \cdot \frac{\omega_0^2+\omega^2}{(\omega_0^2-\omega^2)^2}$$

$$= \frac{4\sigma_1(\omega_0^2+\omega^2)}{(\omega_0^2-\omega^2)^2+4\sigma_1^2\omega^2}$$

and normalised to the value of τ at $\omega = 0$, given by $\tau_0 = 4\sigma_1/\omega_0^2$, we have

$$\frac{\tau(\omega)}{\tau_0} = \frac{(\omega_0^2+\omega^2)\omega_0^2}{(\omega_0^2-\omega^2)^2+4\sigma_1^2\omega^2} \qquad [8.98]$$

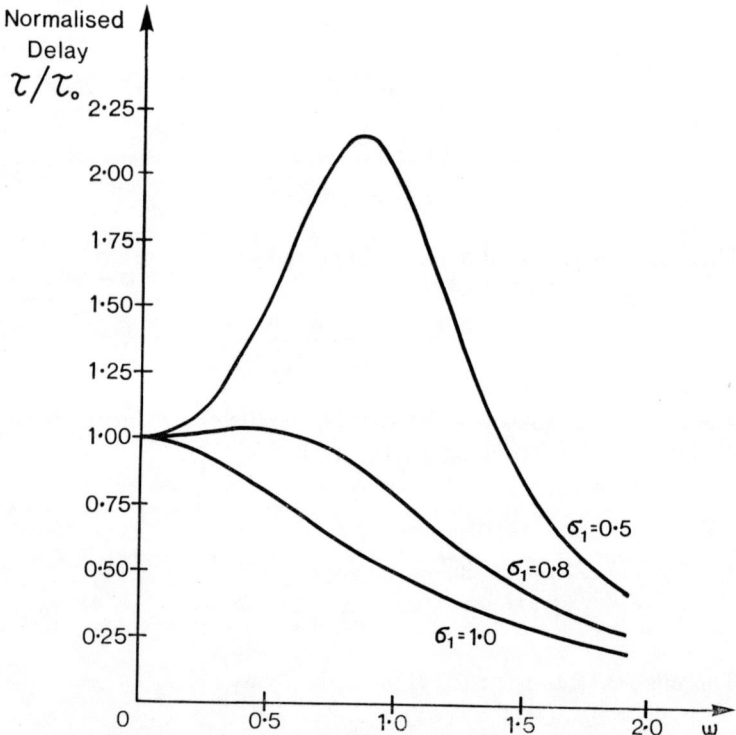

FIG. 8.24. *Normalised group-delay characteristics of all-pass D-sections, $\omega_0 = 1$.*

It is possible to produce a wide variety of group delay characteristics from this all-pass D-section, and some typical characteristics for $\omega_0 = 1$ and $\sigma_1 = 0.5, 0.8$, and 1.0 are plotted in Fig. 8.24. Here we see that it is possible to produce a characteristic with a well defined peak in group delay ($\sigma_1 = 0.5$), one which is almost 'maximally flat' ($\sigma_1 = 0.8$), or one which decreases monotonically ($\sigma_1 = 1.0$). The latter case is rather similar to the C-section characteristic of Fig. 8.22. The D-section has greater flexibility and enables one to achieve much better equalisation of filter phase characteristic than the C-section.

In practice the C- and D-sections are usually realised in forms other than the lattice network we have described, e.g. it is more common to employ the 'twin T' network. The reader will find further information on this in the problems section.

Problems

8.1. Find the insertion loss as a function of frequency of the low pass filter shown in Fig. 8.25, assuming source and load resistances of 1Ω.

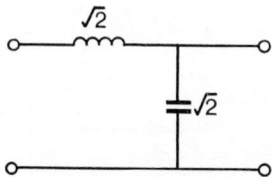

FIG. 8.25. *Problems 8.1, 8.2, 8.9. Element values Henrys and Farads.*

8.2. Using the network of Fig. 8.25 as a prototype, design a band pass filter without transformers with cut-off frequencies (3dB points) at 1 MHz and 2 MHz to work between a source resistance of 100Ω and a load resistance of 2500Ω.

8.3. Assuming source and load resistances of 1Ω, find the insertion loss of the low pass filter shown in Fig. 8.26, and show that the insertion loss has a local minimum at $\omega = 10/\sqrt{7}$ (approximately). What is the value of the insertion loss at this frequency? At what other frequency (to one decimal place) does the insertion loss have this value?

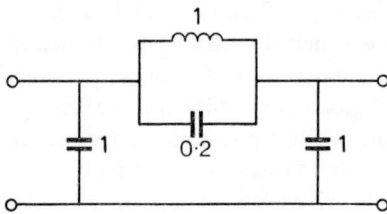

FIG. 8.26. *Problems 8.3 and 8.4. Element values Henrys and Farads.*

8.4. Using the network of Fig. 8.26 as a prototype, design a band stop filter to give an insertion loss of 25·75 dB between frequencies of 10 MHz and 20 MHz, with source and load resistances of 50 Ω.

8.5. Find the transfer impedance of the network of Fig. 8.27, and write down its squared modulus at real frequencies. At which frequencies is this quantity zero? Plot this modulus, and find the frequency between $\omega = 2$ and $\omega = 10$ where it is a minimum, and value of the minimum.

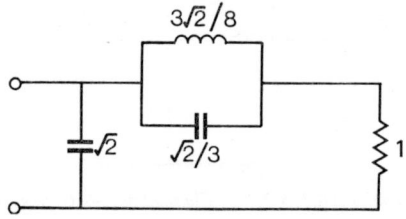

FIG. 8.27. *Problems 8.5 and 8.11. Element values Ohms, Farads, Henrys.*

8.6. Plot the group delay function of [8.98] over the frequency range $0 \leq \omega \leq 16$ for the two cases

(a) $\sigma_1 = 2$, $\omega_0 = 10$

(b) $\sigma_1 = 1$, $\omega_0 = 10$

Write down the approximate peak values and frequency of the delay maximum in each case.

CIRCUIT TRANSFORMATION AND FILTERS 369

8.7. Show that if $\sigma_1/\omega_0 \ll 1$ and $|\omega-\omega_0|/\omega_0 \ll 1$ the delay of the D-section given by [8.98] may be expressed approximately as

$$\tau = \frac{\tau_c}{1+(\Delta\omega/\sigma_1)^2}$$

where τ_c is the group delay at $\omega = \omega_0$.

Write down the expressions for τ_c and $\Delta\omega$. Under what conditions may the D-section be used to equalise the group delay of a narrow band-width filter?

8.8. The group delay of a prototype filter is given by a function $\tau_0(\omega')$. The prototype is converted into a band-pass filter using the transformation

$$\omega' \to Q\left(\frac{\omega}{\omega_0} - \frac{\omega_0}{\omega}\right).$$

Show that the group delay of the band-pass filter is given by

$$\tau(\omega) = Q\left(\frac{1}{\omega_0} + \frac{\omega_0}{\omega^2}\right)\tau_0(\omega')$$

8.9. The low pass prototype of Fig. 8.25 is used to design a bandpass filter having a Q of 10 and mid band frequency at $\omega = 1$. Write down the expression for the group delay of this filter. Give an approximate expression for this delay valid near the frequency $\omega = 1$.

8.10. Find a suitable D-section equaliser to equalise the group delay of the bandpass filter formed in Problem 8.9. The combined group delay of the filter and its equaliser should not vary by more than 0·5 sec over the central 50% of the passband of the filter.

8.11. Find the group delay of the filter shown in Fig. 8.27. Equalise this group delay at $\omega = 0$ and $\omega = 1/\sqrt{2}$ using a C-section equaliser, and find the maximum deviation from a flat group delay response in the band $0 < \omega < 1/\sqrt{2}$ (this occurs approximately where $\omega = 0.525$). What are the circuit element values of the C-section?

370 CIRCUIT THEORY

8.12. By forming the admittance matrix of the twin-T network of Fig. 8.28 and using [*8.65*], or by any other method, derive an equivalent symmetrical lattice network.

Express the conditions for the twin-T to be physically realisable without mutual coupling between the inductors in terms of the circuit elements of the symmetrical lattice network. What does this condition imply in terms of the location of the poles and zeros in the complex plane?

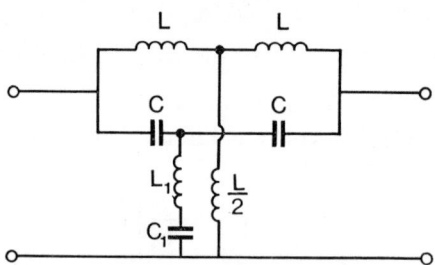

FIG. 8.28. *Problem 8.12.* $L_1 = \left(\dfrac{R^2C}{\alpha}-L\right)\Big/2, \quad C_1 = 2\left(\dfrac{L}{R^2}-C\right)$

$$\alpha = 1-\frac{R^2C}{L}.$$

8.13. Find the symmetrical lattice network equivalent to the bridged-T network shown in Fig. 8.29.

Under what conditions is the symmetrical lattice network realisable as a bridged-T without mutual coupling between the inductors?

How are these conditions modified if mutual coupling is allowed?

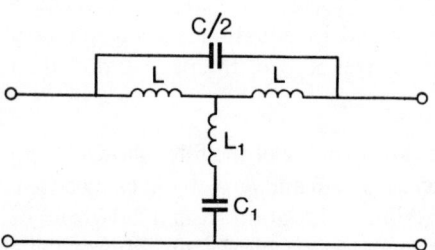

FIG. 8.29. *Problem 8.13.* $L_1 = (R^2C-L)/2, \; C_1 = 2L/R^2$.

Chapter 9

POSITIVE REAL IMPEDANCE FUNCTIONS

In Volume 1 of Circuit Theory and in Chapters 7 and 8 of this book we have been mainly concerned with analysis of networks. This is the situation where we are given the network and the excitation and are required to determine the response. In this chapter the first steps towards the formulation of synthesis procedures will be taken. In synthesis we are given the excitation and the response and are required to find the network. It is with this latter type of problem that engineering is mainly concerned. The area of passive linear networks is one where the synthesis aspect has made great progress and precise mathematical methods exist for accomplishing the task of synthesis.

It should be emphasised that every proposed synthesis problem does not have an answer. For example if one is required to find a linear passive network such that when a voltage $E \sin \omega t$ is applied to a pair of terminals the resulting current is $I_0 \delta(t)$, where $\delta(t)$ is the delta function defined in Chapter 6 of Vol. 1, then one knows from previous knowledge that no solution exists.

Initially we shall be concerned with driving point functions. The prescribed excitation and response will be the current and voltage (or *vice versa*) at a single pair of terminals. The class of networks considered will be linear passive bilateral time invariant *RLC* networks, i.e. the only components allowed will be positive resistors, capacitors and inductors and coupled coils. Before embarking on any specific discussions it is necessary to clarify some rather general points. The precise objective of a synthesis procedure is to find a set of constraints applicable to any prescribed excitation/response specification such that at least one network of the desired class can be found to satisfy the prescription *if and only if* the constraints are satisfied. Such a set of constraints are then said to be *necessary and sufficient*. A set of constraints is *necessary* if they must be satisfied to achieve the objective (but they might be incomplete), while a set of constraints is *sufficient* if success is assured when they are satisfied (but they might be over-restrictive). A set of constraints is both *necessary and sufficient* when they are such that they *cannot* be further relaxed and *need not* be further strengthened in order to assure success.

The necessity of a set of constraints is demonstrated by general analysis of the class of networks required while sufficiency of a set of constraints can only be demonstrated by producing a procedure which generates a network of the required class by relying only on the stated constraints. This chapter is concerned only with necessary conditions while Chapters 10, 11 and 12, are mainly concerned with the question of sufficiency.

As well as establishing the necessary conditions for a driving point impedance we shall explore some consequences of these conditions and present some related material, for later use, on the subject of scattering parameters.

At first sight it would seem a formidable task to formulate conditions applicable to any arbitrary excitation/response pair. However, since we are concerned only with linear networks the situation is much simplified. For example if the excitation is a voltage $E(t)$ and the response is a current $I(t)$ at a single pair of terminals then

$$I(t) = Y(t)E(t) \qquad [9.1]$$

where $Y(t)$ is an admittance operator.

Now since the required network is linear and time-variant it is entirely specified by a knowledge of $Y(t)$. In other words only the ratio of the response to the excitation is significant. Thus for such networks we need only concern ourselves with impedance or admittance functions. Following the method of the Laplace transform given in Chapter 6 of Vol. 1, such immittance (impedance or admittance) functions may be expressed in terms of the complex frequency variable s. Thus referring to driving point functions the problem reduces to finding the necessary and sufficient conditions which a function of the variable s must satisfy if it is to be the driving point impedance or admittance of an *RLC* network. We now proceed to consider the necessary conditions.

9.1. Energy in *RLC* Networks

We consider a general network of l-loops, each loop being coupled to every other loop. If the network contains only *RLC* and mutual

POSITIVE REAL IMPEDANCE FUNCTIONS

inductance then we may write for the jth loop, having an excitation e_j

$$\sum_{k=1}^{l} z_{jk} i_k = e_j \qquad [9.2]$$

where e_j and i_k are functions of time and z_{jk} is an impedance operator

$$z_{jk} = R_{jk} + L_{jk}\frac{d}{dt} + \frac{1}{C_{jk}}\int dt \qquad [9.3]$$

When $j = k$, z_{jj} is the self impedance of loop j while when $j \neq k$ z_{jk} is the mutual impedance between loops j and k. For reciprocal networks $z_{jk} = z_{kj}$. By choosing sufficient loops (including perhaps loops which might otherwise be redundant) it is always possible to write the loop equation in the form of [9.2] with the z_{jk} in the form of [9.3]. The complete set of equations on a loop current basis is therefore

$$\sum_{k=1}^{l} z_{jk} i_k = e_j \qquad j = 1, 2, 3 \ldots n \qquad [9.4]$$

or

$$\sum_{k=1}^{l} \left(R_{jk} i_k + L_{jk}\frac{di_k}{dt} + \frac{1}{C_{jk}}\int i_k dt \right) = e_j \qquad [9.5]$$

The instantaneous power entering the network in loop j is $e_j i_j$ and so the total instantaneous power entering the network is

$$p = \sum_{j=1}^{l} e_j i_j \qquad [9.6]$$

$$= \sum_{j=1}^{l} \sum_{k=1}^{l} \left(R_{jk} i_j i_k + L_{jk} i_j \frac{di_k}{dt} + \frac{1}{C_{jk}} i_j \int i_k dt \right) \qquad [9.7]$$

and putting

$$q = \int i\, dt$$

$$p = \sum_{j=1}^{l} \sum_{k=1}^{l} \left(R_{jk} i_j i_k + L_{jk} i_j \frac{di_k}{dt} + \frac{1}{C_{jk}} q_k \frac{dq_j}{dt} \right) \qquad [9.8]$$

We now define three quantities as follows:

$$F = \sum_{j=1}^{l} \sum_{k=1}^{l} R_{jk} i_j i_k$$

$$T = \tfrac{1}{2} \sum_{j=1}^{l} \sum_{k=1}^{l} L_{jk} i_j i_k$$

$$V = \tfrac{1}{2} \sum_{j=1}^{l} \sum_{k=1}^{l} \frac{1}{C_{jk}} q_j q_k \qquad [9.9]$$

CIRCUIT THEORY

F is the instantaneous dissipated power in the network with a set of loop currents i_j; T is the instantaneous stored energy in the inductors (i.e. the stored magnetic energy) with a set of loop currents i_j; V is the instantaneous stored energy in the capacitors (i.e. the stored electric energy) with a set of loop currents i_j.

For any real physical network, $F, T, V, \geq 0$ for any arbitrary real choice of the set of loop currents i_j. This simple physical fact is the basis upon which we shall derive the necessary conditions related to a driving point impedance.

Let us consider Equations [9.9] in some more detail. F may be written in matrix form as

$$F = [i_1 i_2 i_3 \ldots i_l] \begin{bmatrix} R_{11} & R_{12} \ldots R_{1l} \\ R_{12} & R_{22} \ldots R_{2l} \\ \cdot & \cdot \\ \cdot & \cdot \\ \cdot & \cdot \\ R_{1l} & R_{2l} \ldots R_{ll} \end{bmatrix} \begin{bmatrix} i_1 \\ i_2 \\ \cdot \\ \cdot \\ \cdot \\ i_l \end{bmatrix} \quad [9.10]$$

with similar forms for T and V. We may also write

$$F = (i)^t (R)(i) \quad [9.11]$$

where (i) is the column vector (matrix) of the currents, superscript t indicates transpose, and the square matrix (R) is called the loop resistance matrix. Now we know that for any arbitrary real column vector of currents F is non-negative which implies some particular property of the matrix (R). In fact (R) is said to be the matrix of a positive semi-definite quadratic form, or is said to be positive semi-definite. Similarly the corresponding matrices (L), the loop inductance matrix, and $(1/C)$ the loop elastance matrix, are positive semi-definite.

Returning now to [9.9] let us calculate the time derivative of T.

$$\frac{d}{dt}(T) = \tfrac{1}{2} \sum_{j=1}^{l} \sum_{k=1}^{l} L_{jk} \left(i_j \frac{di_k}{dt} + i_k \frac{di_j}{dt} \right) \quad [9.12]$$

POSITIVE REAL IMPEDANCE FUNCTIONS

But since $L_{jk} = L_{kj}$

$$\frac{dT}{dt} = \tfrac{1}{2}\sum_{j=1}^{l}\sum_{k=1}^{l} L_{jk}i_j \frac{di_k}{dt} + \tfrac{1}{2}\sum_{j=1}^{l}\sum_{k=1}^{l} L_{jk}i_j \frac{di_k}{dt} \qquad [9.13]$$

where i_j and i_k have been interchanged in the second summation since the double summation extends over the same range for j and k. Thus

$$\frac{dT}{dt} = \sum_{j=1}^{l}\sum_{k=1}^{l} L_{jk}i_j \frac{di_k}{dt} \qquad [9.14]$$

similarly

$$\frac{dV}{dt} = \sum_{j=1}^{l}\sum_{k=1}^{l} \frac{1}{C_{jk}} q_k \frac{dq_j}{dt} \qquad [9.15]$$

and so

$$P = F + \frac{d}{dt}(T+V) \qquad [9.16]$$

which merely expresses the fact that the power entering the network is the sum of the dissipated power and the rate of change of the stored energy.

The main object of this section has been to establish the fact that the matrices (R), (L) and $(1/C)$ are positive semi-definite.

A similar dual analysis on a node-pair voltage basis could obviously be carried out. The excitations in that case would be a set of currents i_j and corresponding to [9.2] we find

$$\sum_{k=1}^{l} y_{jk}e_k = i_j \qquad j = 1, 2, 3\ldots n \qquad [9.17]$$

with

$$y_{jk} = \hat{G}_{jk} + \hat{C}_{jk}\frac{d}{dt} + \frac{1}{\hat{L}_{jk}}\int dt \qquad [9.18]$$

In this case

$$\hat{F} = \sum_{j=1}^{n}\sum_{k=1}^{n} \hat{G}_{jk}e_je_k$$

$$\hat{T} = \tfrac{1}{2}\sum_{j=1}^{n}\sum_{k=1}^{n} \frac{1}{\hat{L}_{jk}} x_j x_k \quad \text{where} \quad x = \int e\,dt$$

$$\hat{V} = \tfrac{1}{2}\sum_{j=1}^{n}\sum_{k=1}^{n} \hat{C}_{jk}e_je_k \qquad [9.19]$$

where $\hat{F}$ is the instantaneous dissipated power and $\hat{T}$ and $\hat{V}$ are the instantaneous stored magnetic and electric energy respectively. Again by the same arguments as before, the matrices $(\hat{G})$, $(1/\hat{L})$ and $(\hat{C})$ are positive semi-definite. Also

$$p = \sum_{j=1}^{n} e_j i_j = \hat{F} + \frac{d}{dt}(\hat{T} + \hat{V}) \qquad [9.20]$$

It should be noted that in either case if there is only a single excitation (say, e_r or i_r) then $p = e_r i_r$.

9.2. Properties of Driving Point Impedances

We now apply the Laplace transformation to Equation [9.2] to obtain

$$\sum_{k=1}^{l} Z_{jk} I_k = E_j \qquad j = 1, 2, \ldots l \qquad [9.21]$$

where now E_j and I_k are functions of the Laplace transform variable (or complex frequency) s and are in fact the Laplace transforms of i_k and e_j. Now

$$Z_{jk} = R_{jk} + L_{jk} s + \frac{1}{C_{jk} s} \qquad [9.22]$$

and we have assumed quiescent initial conditions, i.e. $i_k(0) = 0$ and $\int_{-\infty}^{0} i_k dt = 0$. This does not constitute a restriction since we know from Chapter 6 of Vol. 1, that such terms may be considered as sources which can be incorporated in E_j whose form we have not had to specify.

We now wish to consider the expression

$$\sum_{j=1}^{l} E_j I_j^* = \sum_{j=1}^{l} \sum_{k=1}^{l} \left(R_{jk} I_k I_j^* + L_{jk} s I_k I_j^* + \frac{1}{C_{jk} s} I_k I_j^* \right) \qquad [9.23]$$

and we define the following quantities

$$F_0 = \sum_{j=1}^{l} \sum_{k=1}^{l} R_{jk} I_k I_j^*$$

$$T_0 = \sum_{j=1}^{l} \sum_{k=1}^{l} L_{jk} I_k I_j^*$$

$$V_0 = \sum_{j=1}^{l} \sum_{k=1}^{l} \frac{1}{C_{jk}} I_k I_j^* \qquad [9.24]$$

so that
$$\sum_{j=1}^{l} E_j I_j^* = F_0 + sT_0 + \frac{V_0}{s} \qquad [9.25]$$

Now we may also write

$$F_0 = [(I_1 I_2 \ldots I_l)(R)] \begin{bmatrix} I_1^* \\ I_2^* \\ \cdot \\ \cdot \\ \cdot \\ \cdot \\ I_l^* \end{bmatrix} \qquad [9.26]$$

$$T_0 = [(I_1 I_2 \ldots I_l)(L)] \begin{bmatrix} I_1^* \\ I_2^* \\ \cdot \\ \cdot \\ \cdot \\ \cdot \\ I_l^* \end{bmatrix} \qquad [9.27]$$

$$V_0 = [(I_1 I_2 \ldots I_l)(1/C)] \begin{bmatrix} I_1^* \\ I_2^* \\ \cdot \\ \cdot \\ \cdot \\ \cdot \\ I_l^* \end{bmatrix} \qquad [9.28]$$

or
$$F_0 = (I)(R)(\check{I})$$
$$T_0 = (I)(L)(\check{I})$$
$$V_0 = (I)(1/C)(\check{I}) \qquad [9.29]$$

Where I is a row vector of the currents and the symbol $(\check{I})$ denotes complex conjugate transpose. Now in this case although (R), (L) and $(1/C)$ are still the same positive semi-definite matrices the vector (I) has (in general) elements which are complex and we must discover the properties of F_0, T_0, and V_0, which do not have the physical significance of F, T and V.

First it is easy to see that F_0, T_0 and V_0 are real. For example if we write

$$F_0 = \sum_{j=1}^{l} \sum_{k=1}^{l} R_{jk} I_j I_k^*$$

then

$$F_0^* = \sum_{j=1}^{l} \sum_{k=1}^{l} R_{jk} I_j^* I_k$$

and since the subscripts j and k indicate the same range of summation

$$F_0^* = F_0 \qquad [9.30]$$

so that F_0 must be real for any arbitrary (complex) set of currents. Similarly T_0 and V_0 are real.

Now if we write
$$(I) = (\alpha) + j(\beta)$$
i.e.
$$(I_1 I_2 \ldots I_l) = (\alpha_1 + j\beta_1 \ \alpha_2 + j\beta_2 \ldots \alpha_l + j\beta_l)$$
then
$$F_0 = [(\alpha) + j(\beta)](R)[(\alpha)^t - j(\beta)^t]$$
$$= [(\alpha)(R)(\alpha)^t + (\beta)(R)(\beta)^t] + j[(\beta)(R)(\alpha)^t - (\alpha)(R)(\beta)^t]$$

The imaginary part is, of course, zero since the second term is merely the transpose of the first which is itself a single number. Thus

$$F_0 = (\alpha)(R)(\alpha)^t + (\beta)(R)(\beta)^t \qquad [9.31]$$

where (α) and (β) are arbitrary real row vectors. But we know from the previous section that since (R) is positive semi-definite an expression such as $(x)(R)(x)^t$ is non-negative for any real row vector (x). Thus F_0 being the sum of two such expressions is non-negative for any arbitrary row vectors (α) and (β). Hence F_0 is real and non-negative for any arbitrary set of (complex) currents (I). Similarly T_0 and V_0 are real and non-negative. Thus we obtain the key result that from [9.25]

$$\sum_{j=1}^{l} E_j I_j^* = F_0 + sT_0 + (V_0/s) \qquad [9.32]$$

where F_0, T_0 and V_0 are real and non-negative for any arbitrary set of (complex) currents I_j. The quantities F_0, T_0 and V_0 are known as the 'Energy Functions' of the network.

If we now consider [9.32] in the case where there is only a single excitation in loop r, say, then we obtain

$$E_r I_r^* = F_0 + sT_0 + (V_0/s) \qquad [9.33]$$

or

$$\frac{E_r}{I_r} = \frac{1}{I_r I_r^*}\left(F_0 + sT_0 + \frac{V_0}{s}\right) = Z_{rr} \qquad [9.34]$$

In other words the driving point impedance seen by the generator in loop r can be expressed in the form given by [9.34]. This situation is quite general and any arbitrary driving point impedance can be expressed in this way. Now the quantity $I_r I_r^*$ is a real number and its magnitude may be set to any arbitrary value. In other words it is merely a scaling factor on the impedance.

By exactly similar means an analysis on a node-pair voltage basis could be carried out. Definining $\hat{F}_0$, $\hat{T}_0$, and $\hat{V}_0$ from the form of [9.19] it is shown by the dual analysis that these are real and non-negative and that with excitation I_r at node r

$$\frac{I_r}{E_r} = \frac{1}{E_r E_r^*}\left(\hat{F}_0 + \frac{\hat{T}_0}{s} + s\hat{V}_0\right) = Y_{rr} \qquad [9.35]$$

All the necessary conditions on a driving point impedance may be found with the aid of [9.34] and [9.35]. First of all it is clear that any impedance is a polynomial function of s, since in its calculation we only use sums and products of terms like R_{jk}, $L_{jk}s$, $C_{jk}s$. In fact any impedance must be, in general, the ratio of two polynomials in this variable. Now from [9.34] it is evident that Z is real when s is made real since F_0, T_0 and V_0 are all real. Thus Z is the ratio of two polynomials in s with real coefficients.

If, in general, we put $s = \sigma + j\omega$ then

$$Z_{rr} = \frac{1}{I_r I_r^*}\left[F_0 + \sigma T_0 + \frac{\sigma V_0}{\sigma^2 + \omega^2} + j\left(\omega T_0 - \frac{\omega V_0}{\sigma^2 + \omega^2}\right)\right] \quad [9.36]$$

and we can conclude that

$$\operatorname{Re} Z > 0 \quad \text{for} \quad \sigma > 0 \quad [9.37]$$

Firstly $\operatorname{Re} Z$ cannot be negative since F_0, T_0, V_0 are non-negative and if $\operatorname{Re} Z = 0$ it would require F_0, T_0, and V_0 to be simultaneously zero. But this indicates that for F_0 to be zero there should be no current in any resistor, for T_0 to be zero there should be no current in any inductor and for V_0 to be zero there should be no current in any capacitor. This can only occur in a null network since I_r is assumed finite. Thus [9.37] is established. Note, however, that it is quite possible for any two of F_0, T_0 and V_0 to be zero.

In general, therefore, we see that any impedance $Z(s)$ is such that

$Z(s)$ is real for s real
$\operatorname{Re} Z(s) > 0 \quad \text{for} \quad \operatorname{Re} s > 0$

such a function is said to be *positive real (p.r.)*

The positive real condition implies many other interesting and useful properties which we may call upon in developing synthesis procedures.

9.2.1. CONSEQUENCES OF THE POSITIVE REAL CONDITION

(i) *If a function is positive real its reciprocal is also positive real.* It is clear that if $Z(s)$ is real for s real then $1/Z(s)$ is also real for s real. Now if $Z(s) = R(s) + jX(s)$ and $R(s) > 0$ for $\operatorname{Re} s > 0$ then

$$\frac{1}{Z(s)} = \frac{R(s) - jX(s)}{R^2(s) + X^2(s)}$$

and

$$\operatorname{Re}\{1/Z(s)\} = \frac{R(s)}{R^2(s) + X^2(s)}$$

so

$$\operatorname{Re}\{1/Z(s)\} > 0 \quad \text{if} \quad R(s) > 0 \quad \text{i.e. for} \quad \operatorname{Re} s > 0.$$

(ii) *A positive real function has no poles or zeros in the right half s-plane* ($\operatorname{Re} s > 0$). Clearly there are no zeros since $\operatorname{Re} Z(s) > 0$ throughout $\operatorname{Re} s > 0$. Any poles of $Z(s)$ are zeros of $Y(s) = 1/Z(s)$.

But $Y(s)$ is positive real if $Z(s)$ is positive real hence $Y(s)$ has no zeros and therefore $Z(s)$ has no poles in Re $s > 0$.

(iii) *Any poles of $Z(s)$ or $Y(s)$ on the imaginary axis $(s = j\omega)$ are simple with real positive residues.* Let s_0 be a pole of $Z(s)$ of order n, located on the imaginary axis. $Z(s)$ may be expanded in a Laurent series about this pole, and for s in the vicinity of s_0, $Z(s)$ may be approximated by the dominant term in the expansion. Thus for s in the vicinity of s_0 we may write

$$Z(s) \approx \frac{a_{-n}}{(s-s_0)^n}$$

where a_{-n} is the residue.

Now let
$$a_{-n} = \rho \exp(j\theta)$$
$$s-s_0 = r \exp(j\phi)$$

so that
$$Z(s) = \rho \exp\{j(\theta-n\phi)\}/r^n$$

and
$$\text{Re } Z(s) = \rho \cos(\theta-n\phi)/r^n \qquad [9.38]$$

Referring now to Fig. 9.1 let us assume that the point s describes an arc of a circle in the right half plane with $|s-s_0|$ as small as we please

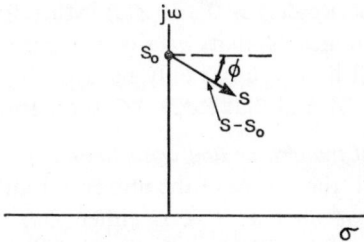

FIG. 9.1. *Illustration of a positive real function in the vicinity of an imaginary axis pole.*

but not zero. Let there be a value of ϕ (say ϕ_1) such that $\cos(\theta-n\phi)$ is positive (Re $Z(s)$ positive) then if ϕ is increased by an amount π/n the argument of $\cos(\theta-n\phi)$ will change by an amount π and Re $Z(s)$ will become negative. But if $n \geq 2$, ϕ must vary by at least π/n in the right half plane and Re $Z(s)$ becomes negative at some stage. But if $Z(s)$ is positive real this cannot occur so n can at most be one.

Thus any pole must be simple. Now in the right half plane ϕ varies between $-(\pi/2)$ and $\pi/2$ so that unless $\theta = 0$, $\cos(\theta - \phi)$ will again become negative. $\theta = 0$ implies that a_{-n} is real and positive. Thus a pole on the imaginary axis must be simple ($n = 1$) and have a real positive residue ($\theta = 0$). Similarly poles of $Y(s)$ (zeros of $Z(s)$) must be simple and possess a real positive residue.

(iv) *If $Z(s)$ is positive real then* $\operatorname{Re} Z(s) \geq 0$ *for* $\operatorname{Re} s = 0$. Let $Z(s) = Z(\sigma, j\omega)$ for $s = \sigma + j\omega$ and since we are dealing with polynomial functions $\operatorname{Re} Z(\sigma, j\omega)$ is a continuous function of σ for any given value of ω, now we know that $\operatorname{Re} Z(\varepsilon, j\omega_1) > 0$ for any positive ε however small. Suppose that $\operatorname{Re} Z(0, j\omega_1)$ is negative then for some value of σ, $0 < \sigma \leq \varepsilon$, $\operatorname{Re} Z(\sigma, j\omega_1) = 0$ which would violate the condition that $\operatorname{Re} Z(\sigma, j\omega_1) > 0$ for $\sigma > 0$. It is, of course, possible for $\operatorname{Re} Z(0, j\omega_1) = 0$ since this would not violate the requirement $\operatorname{Re} Z(\sigma, j\omega_1) > 0$ for $\sigma > 0$.

(v) *If $Z(s)$ is real for s real, has no poles in the right half plane, and $\operatorname{Re} Z(s) \geq 0$ for $\operatorname{Re} s = 0$, then $Z(s)$ is positive real*. Since $Z(s)$ has no poles in the right half plane it is analytic there and the minimum real part theorem states that under these conditions the minimum value of the real part occurs on the boundary (in this case the imaginary axis). But the minimum value of $\operatorname{Re} Z(s)$ for $\operatorname{Re} s = 0$ is zero so throughout the right half plane $\operatorname{Re} Z(s) > 0$ and $Z(s)$ is therefore positive real.

This property indicates that we do not need to test whether $\operatorname{Re} Z(s) > 0$ for all $\operatorname{Re} s > 0$ but only need to show the absence of right half plane poles and that $\operatorname{Re} Z(s) \geq 0$ for $\operatorname{Re} s = 0$.

(vi) *The degrees of numerator and denominator of $Z(s)$ may differ at most by one.* If η_N is the degree of the numerator and η_D is the degree of the denominator then as $s \to \infty$ either $Z(s)$ or $Y(s)$ tends to $s^{|\eta_N - \eta_D|}$ giving a pole of order $|\eta_N - \eta_D|$ at infinity. But the point at infinity is on the imaginary axis and if $Z(s)$ is positive real the order of this pole cannot exceed unity. Therefore $|\eta_N - \eta_D| \leq 1$.

(vii) *If $Z(s)$ is positive real then*

$$Z_1(s) = Z(s) - \frac{2\alpha s}{s^2 + \omega_i^2}$$

is positive real if and only if $Z(s)$ has poles at $s = \pm j\omega_i$ with residue $k_i \geq \alpha$. If $Z(s)$ does not have poles at $s = \pm j\omega_i$, $Z_1(s)$ has such poles with residue $-\alpha(<0)$ and so is not positive real. If $Z(s)$ has such poles with residue k_i then carrying out a partial fraction expansion of $Z(s)$ yields

$$Z(s) = \frac{2k_i s}{s^2 + \omega_i^2} + Z_r(s)$$

and

$$Z_1(s) = \frac{2(k_i - \alpha)s}{s^2 + \omega_i^2} + Z_r(s)$$

Now Z_r has no right half plane poles since poles of Z_r are poles of Z and $\operatorname{Re} Z_r(j\omega) \geq 0$ since $\operatorname{Re} Z_r(j\omega) = \operatorname{Re} Z(j\omega)$. Thus $Z_r(s)$ is positive real. Now $Z_1(s)$ is therefore positive real if, and only if, $k_1 - \alpha \geq 0$, since otherwise there is a negative residue at the poles $s = \pm j\omega_i$.

With the aid of the properties given in this section we are now in a position to examine any given function of s to see if it satisfies the positive real condition i.e. to see if it could possibly represent the driving point impedance or admittance of an *RLC* network. Even if a given function is positive real we do not know whether this is in fact sufficient. However, it will be shown in a later chapter that the positive real condition is indeed sufficient.

9.3. The Bounded Real Condition

A function $\Gamma(s)$ is bounded real if

1. It is real for s real.
2. $0 \leq |\Gamma(s)| < 1$ for $\operatorname{Re} s > 0$.

Obviously $\Gamma(s)$ must be devoid of poles in $\operatorname{Re} s > 0$ and by an argument similar to that given in Condition (iv) of Section 9.2.1 $|\Gamma(s)| \leq 1$ for $\operatorname{Re} s = 0$. Similarly if $\Gamma(s)$ is real for s real and is devoid of right half plane poles with $|\Gamma(s)| \leq 1$ for $\operatorname{Re} s = 0$ this is equivalent to the bounded real condition as stated above.

If $Z(s)$ is positive real then

$$\Gamma(s) = \pm \frac{Z(s) - 1}{Z(s) + 1} \qquad [9.39]$$

is bounded real.

Clearly $\Gamma(s)$ is real for s real. Now for $\operatorname{Re} s > 0$, let $Z(s) = R + jX$

and $R > 0$ in this domain by virtue of the positive real condition. Also

$$\Gamma = \pm\frac{R+jX-1}{R+jX+1}$$

and

$$|\Gamma|^2 = \frac{(R-1)^2+X^2}{(R+1)^2+X^2} \geq 0$$

$$= 1-\frac{4R}{(R+1)^2+X^2} \qquad [9.40]$$

Thus if $R > 0$, $|\Gamma| < 1$ and so $\Gamma(s)$ is bounded real. Also, it is clear from [9.39] that $\Gamma(s)$ is devoid of right half plane poles (since $Z(s)$ would have to take the value -1 to produce such a pole) and from [9.40] $|\Gamma|^2 \leq 1$ for $R \geq 0$ i.e. for $\mathrm{Re}\, s = 0$.

The converse of [9.30] is also true. Thus: if $\Gamma(s)$ is bounded real then

$$Z(s) = \frac{1\pm\Gamma(s)}{1\mp\Gamma(s)} \qquad [9.41]$$

is positive real. The reader should prove this as an exercise.

9.4. Hurwitz Polynomials and Reactance Functions

A polynomial in s is called *Hurwitz* if it is devoid of zeros in the right half plane. It follows that a positive real function (for *RLC* networks) is the ratio of two Hurwitz polynomials. Note however that not all ratios of Hurwitz polynomials yield positive real functions. (One must add the condition $\mathrm{Re}\, Z \geq 0$ for $\mathrm{Re}\, s = 0$).

If $Z(s)$ is positive real and $Z(s)+Z(-s) \equiv 0$ then $Z(s)$ is called a *reactance function* or a *Foster function*. A reactance function must be the ratio of an even to an odd polynomial or vice versa, since it is an odd function. Explicitly, if we write

$$Z(s) = \frac{m_1+n_1}{m_2+n_2} \qquad [9.42]$$

where m_1 and n_1 are the even and odd parts of the numerator polynomial respectively, and m_2 and n_2 are the corresponding parts of the denominator polynomial, then

$$Z(s)+Z(-s) = \frac{m_1+n_1}{m_2+n_2}+\frac{m_1-n_1}{m_2-n_2}$$

$$= \frac{2(m_1m_2-n_1n_2)}{m_2^2-n_2^2} \equiv 0$$

Hence

$$\frac{m_1}{n_1} \equiv \frac{n_2}{m_2}$$

and

$$Z(s) = \frac{n_1}{m_2} \qquad [9.43]$$

or

$$Z(s) = \frac{m_1}{n_2} \qquad [9.44]$$

Now since $Z(s)$ is positive real it cannot have any right half plane poles or zeros. Similarly $Z(-s)$ cannot have any left half plane poles or zeros. But $Z(s) = -Z(-s)$ so that $Z(s)$ cannot have either right or left half plane poles or zeros. Thus *the poles and zeros of a reactance function lie entirely on the imaginary axis.* We shall see many further properties of reactance functions in the next chapter.

If $Z(s)$ is a reactance function $= m/n$ or $= n/m$ then $m+n$ is a Hurwitz polynomial. $Z(s)$ is positive real hence $Z(s)+1$ is clearly positive real. But $Z(s)+1 = (m+n)/n$ or $(m+n)/m$. Now we know that a positive real function is the ratio of two Hurwitz polynomials so therefore $m+n$ is a Hurwitz polynomial.

If $m+n$ is a Hurwitz polynomial then $Z(s) = m/n$ is a reactance function. Clearly $Z(s)+Z(-s) \equiv 0$ and we need only to show that $Z(s)$ is positive real. $Z(s)$ is real for s real and if we form

$$\Gamma(s) = \pm\frac{1-Z(s)}{1+Z(s)} = \pm\frac{n-m}{n+m}$$

we see that $\Gamma(s)$ has no right half plane poles since $m+n$ is a Hurwitz polynomial. Further when $\operatorname{Re} s = 0$, m is real and n is imaginary by virtue of their even and odd nature respectively. Let $n = jN$ and $m = M$ for $s = j\omega$, so that

$$|\Gamma(j\omega)|^2 = \frac{M^2+N^2}{M^2+N^2} \equiv 1$$

Now since $\Gamma(s)$ has no right half plane poles and $|\Gamma(j\omega)| \leq 1$, $Z(s)$ is positive real and is a reactance function.

We shall see that this property of Hurwitz polynomials forms an excellent basis for testing any function to see if it is Hurwitz.

9.5. Two-port Scattering Parameters

In Chapter 5 of Vol. 1, the various two-port parameters were examined in some detail. We shall now introduce a new set of parameters called the *scattering parameters*. These parameters can be used in the description of networks with more than two ports and have important and subtle properties whose implications are beyond the scope of this book. They will be introduced here merely in the context of two-ports and in order to be able to call upon them in a later synthesis procedure.

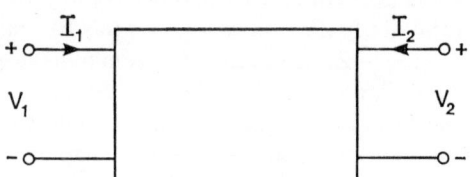

FIG. 9.2. *A two-port network.*

We consider a two-port as shown in Fig. 9.2. If the matrix

$$(V) = \begin{pmatrix} V_1 \\ V_2 \end{pmatrix} \quad \text{and} \quad (I) = \begin{pmatrix} I_1 \\ I_2 \end{pmatrix}$$

then we construct two new matrices as follows:

$$(a) = \begin{pmatrix} a_1 \\ a_2 \end{pmatrix} = \begin{pmatrix} R_1^{-\frac{1}{2}} & 0 \\ 0 & R_2^{-\frac{1}{2}} \end{pmatrix}(V) + \begin{pmatrix} R_1^{\frac{1}{2}} & 0 \\ 0 & R_2^{\frac{1}{2}} \end{pmatrix}(I)$$

$$(b) = \begin{pmatrix} b_1 \\ b_2 \end{pmatrix} = \begin{pmatrix} R_1^{-\frac{1}{2}} & 0 \\ 0 & R_2^{-\frac{1}{2}} \end{pmatrix}(V) - \begin{pmatrix} R_1^{\frac{1}{2}} & 0 \\ 0 & R_2^{\frac{1}{2}} \end{pmatrix}(I) \quad [9.45]$$

(a) and (b) are clearly linear combinations of (V) and (I) and the quantities R_1 and R_2 are arbitrary positive numbers with the dimensions of resistance. They are called the *port normalizing numbers*. Just as the two-port may be described by a relationship between (V) and (I), so it may be described by a relationship between (b) and (a). In fact we write

$$(b) = (S)(a) = \begin{pmatrix} S_{11} & S_{12} \\ S_{21} & S_{22} \end{pmatrix} (a) \qquad [9.46]$$

and (S) is called the scattering matrix of the two-port.

From [9.45] we may obtain (V) and (I) in terms of (a) and (b) to give

$$(V) = \begin{pmatrix} R_1^{\frac{1}{2}} & 0 \\ 0 & R_2^{\frac{1}{2}} \end{pmatrix} [(a)+(b)]$$

$$(I) = \begin{pmatrix} R_1^{-\frac{1}{2}} & 0 \\ 0 & R_2^{-\frac{1}{2}} \end{pmatrix} [(a)-(b)] \qquad [9.47]$$

But

$$(V) = (Z)(I)$$

$$\begin{pmatrix} R_1^{\frac{1}{2}} & 0 \\ 0 & R_2^{\frac{1}{2}} \end{pmatrix} [(a)+(b)] = (Z) \begin{pmatrix} R_1^{-\frac{1}{2}} & 0 \\ 0 & R_2^{-\frac{1}{2}} \end{pmatrix} [(a)-(b)]$$

Substituting $(b) = (S)(a)$ and rearranging

$$[1+(S)](a) = \begin{pmatrix} R_1^{-\frac{1}{2}} & 0 \\ 0 & R_2^{-\frac{1}{2}} \end{pmatrix} (Z) \begin{pmatrix} R_1^{-\frac{1}{2}} & 0 \\ 0 & R_2^{-\frac{1}{2}} \end{pmatrix} [1-(S)](a)$$

Thus

$$(Z) = \begin{pmatrix} R_1^{\frac{1}{2}} & 0 \\ 0 & R_2^{\frac{1}{2}} \end{pmatrix} [1+(S)][1-(S)]^{-1} \begin{pmatrix} R_1^{\frac{1}{2}} & 0 \\ 0 & R_2^{\frac{1}{2}} \end{pmatrix} \qquad [9.48]$$

Similarly (Y) may be expressed in terms of (S) and (S) may be expressed in terms of (Z) or (Y).

We now consider the situation shown in Fig. 9.3. The transducer power gain is given by:

$$G_T(j\omega) = \frac{4|I_2|^2 R_2 R_1}{|E|^2} \qquad [9.49]$$

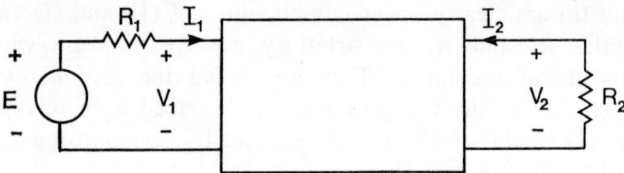

FIG. 9.3. *A resistively terminated two-port.*

If we now choose R_1 and R_2 for the port normalizing numbers then from [*9.47*]

$$I_1 = R_1^{-\frac{1}{2}}(a_1 - b_1)$$
$$I_2 = R_2^{-\frac{1}{2}}(a_2 - b_2)$$
$$V_1 = R_1^{\frac{1}{2}}(a_1 + b_1)$$
$$V_2 = R_2^{\frac{1}{2}}(a_2 + b_2) \qquad [9.50]$$

But
$$V_2 = -R_2 I_2$$
hence
$$a_2 = 0.$$

Also
$$E = R_1 I_1 + V_1 = 2R_1^{\frac{1}{2}} a_1$$

Hence
$$G_T(\omega) = \frac{|b_2|^2}{|a_1|^2} \qquad [9.51]$$

But since $a_2 = 0$, $\quad b_2 = S_{21} a_1$ so that

$$G_T(\omega) = |S_{21}(j\omega)|^2 \qquad [9.52]$$

For a reciprocal network since $Z_{21} = Z_{12}$ we have from [*9.48*] $S_{12} = S_{21}$. [*9.52*] indicates a very important property of the scattering matrix namely that if the terminating resistances are taken as the port normalizing numbers the modulus squared of the diagonal term is the transducer power gain.

Further if we write the input impedance of the network at port 1 we obtain

$$Z_{in} = \frac{V_1}{I_1} = R_1 \frac{a_1 + b_1}{a_1 - b_1} \qquad [9.53]$$

POSITIVE REAL IMPEDANCE FUNCTIONS

But
$$b_1 = S_{11}a_1$$
since
$$a_2 = 0$$
Thus
$$\frac{Z_{in}}{R_1} = \frac{1+S_{11}}{1-S_{11}} \qquad [9.54]$$

Note that since Z_{in} must be positive real for an *RLC* network S_{11} must be bounded real from [9.39] if we write [9.54] at complex frequencies. Similarly by interchanging ports 1 and 2 we may show that S_{22} is also bounded real.

From [9.52] we know that $|S_{12}(j\omega)|^2$ is the transducer power gain at real frequencies and so

$$0 \leq |S_{12}(j\omega)|^2 \leq 1 \qquad [9.55]$$

since the power delivered to the load cannot exceed the available power from the source when the network is passive.

Consider now the power entering the two-port at real frequencies.

$$P = \text{Re}(V_1 I_1^* + V_2 I_2^*)$$
$$= \text{Re}\,[(a_1+b_1)(a_1^*-b_1^*) - b_2 b_2^*]$$

from [9.50] and remembering that $a_2 = 0$. Hence

$$P = a_1 a_1^* - b_1 b_1^* - b_2 b_2^*$$

But
$$b_1 = S_{11}a_1; \qquad b_2 = S_{12}a_2$$
so that
$$P = a_1 a_1^*(1 - S_{11}S_{11}^* - S_{12}S_{12}^*)$$

If the two-port is *lossless* $P \equiv 0$ so that

$$S_{11}S_{11}^* + S_{12}S_{12}^* = 1 \qquad [9.56]$$

This important relationship will be used extensively in later chapters.

9.6. Additional Remarks

The main result of this chapter has been to establish the necessity of the positive real condition for the driving point impedance of *RLC* networks, and to explore some of its consequences. The introduction of the bounded real condition is a valuable aid in manipulating impedance functions. The related concepts of Hurwitz Polynomials and Reactance functions will be seen to have considerable importance in later chapters. The brief outline of the scattering parameter theory for two-ports is an essential tool for use in the chapter on transfer function synthesis. In general the material of this chapter provides a background for much of what follows and a thorough understanding of it is essential as a proper foundation for any study in network synthesis.

Problems

9.1. A network consists of three inductors L_1, L_2, L_3 with mutual couplings L_{12}, L_{23}, L_{31} respectively. Using the fact that the stored magnetic energy is non-negative show that

$$L_1, L_2, L_3, > 0$$
$$L_1 L_2 \geq L_{12}^2$$
$$L_2 L_3 \geq L_{23}^2$$
$$L_3 L_1 \geq L_{31}^2$$
$$L_1 L_2 L_3 \geq L_{12}^2 L_3 + L_{23}^2 L_1 + L_{31}^2 L_2 - 2 L_{12} L_{23} L_{31}$$

9.2. Show by direct calculation from the circuit of Fig. 9.4 that the matrices (R), (L) and $(1/C)$ are positive definite.

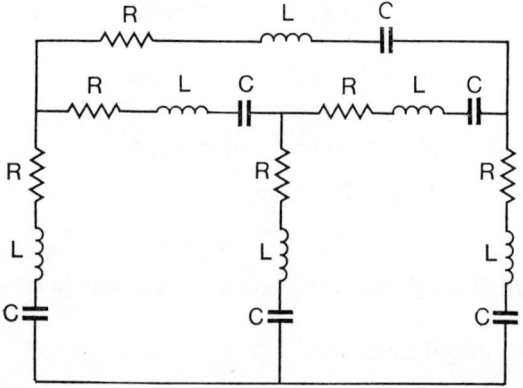

FIG. 9.4. *Problem 9.2.*

9.3. Show that a symmetric matrix where the off-diagonal terms are negative and the sum of each row or column is non-negative, is positive semi-definite. Show also that in an arbitrary *RLC* network without ideal transformers or coupled coils the matrices $(\hat{G})$, $(\hat{C})$ and $(1/\hat{L})$ have the properties listed above.

9.4. If a function $Z(s)$ is positive real discuss the question of whether the following functions are positive real.

(a) $Z(s) + K$ where K is a positive constant.

(b) $Z(s) - K$

(c) $KZ(s)$

(d) $F\{Z(s)\}$ where $F(s)$ is also positive real.

(e) $Z(s) + s$

(f) $Z(s) - s$

(g) $\dfrac{Z(s) + s}{1 + s^2 + sZ(s)}$

9.5. Referring to [9.41] show that
$$Z(s) = \frac{1 \pm \Gamma(s)}{1 \mp \Gamma(s)}$$
is positive real if $\Gamma(s)$ is bounded real.

9.6. If $\Gamma(s)$ is a bounded real function discuss the question of whether the following functions are bounded real.

(a) $\Gamma_1(s) = \dfrac{K + (2 - K)\Gamma(s)}{2 + K - K\Gamma(s)}$ where K is a positive constant

(b) $\Gamma_1(s) = \dfrac{s + (2 - s)\Gamma(s)}{2 + s - s\Gamma(s)}$

(c) $\Gamma_1(s) = \dfrac{-s + (2 + s)\Gamma(s)}{2 - s + s\Gamma(s)}$

(d) $\Gamma_1(s) = \dfrac{(1 - s)\Gamma(s) + 1 + s}{-(1 - s + s^2)\Gamma(s) + (1 + s + s^2)}$

9.7. Check the following functions for positive reality.

(a) $\dfrac{s^2+7s+12}{s^2+3s+2}$

(b) $\dfrac{s^2+7s+12}{s^2+s-2}$

(c) $\dfrac{1}{s^2+3s+2}$

(d) $\dfrac{s^2}{s^3+3s+2}$

9.8. Derive expressions for the scattering matrix of a two-port in terms of (a) the z-matrix (b) the y-matrix, with port normalizing numbers R_1 and R_2.

9.9. Calculate the scattering matrix of the two-port shown in Fig. 9.5 using 1-ohm port normalizing numbers, and hence show that the transducer power gain is $1/(1+\omega^6)$.

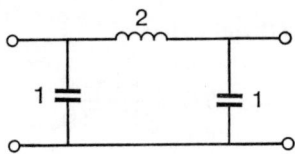

Fig. 9.5. *Problem 9.9.*

9.10. An ideal transformer with turns ratio $n:1$ does not possess either a z- or a y-matrix (i.e. all entries in these matrices are infinite). Calculate the scattering matrix of such a network with 1-ohm port normalizing numbers.

Chapter 10

TWO-ELEMENT-KIND SYNTHESIS

In the previous chapter the necessity of the positive real condition for the driving point impedance of any *RLC* network was established. Logically one might proceed now to try to demonstrate its sufficiency by formulating a procedure to generate suitable networks for this general class. However, in many important applications only two kinds of element may be permitted *RL*, *RC* or *LC*. This chapter is devoted to a discussion of this situation. The approach for two-element-kind networks is rather different to that needed in the general case and in many respects somewhat simpler. We begin by discussing *LC* (or lossless) networks.

10.1. Driving Point Impedance of *LC* Networks

The driving point impedance of an *LC* network must, of course, be positive real since such networks belong to the *RLC* class. The characteristic of such networks is that all resistance disappears. Hence

$$F_0 \equiv 0 \qquad [10.1]$$

and we may write from [9.34]

$$Z_{rr} = \frac{1}{I_r I_r^*}\left(sT_0 + \frac{V_0}{s}\right) \qquad [10.2]$$

or normalizing $I_r I_r^*$ to unity and writing Z in general rather than Z_{rr}

$$Z = sT_0 + (V_0/s) \qquad [10.3]$$

now

$$Z(s) + Z(-s) \equiv 0$$

so that Z is a reactance function as defined in Section 9.4.

Thus Z must be the ratio of an odd to an even polynomial or vice-versa and its poles and zeros must lie on the imaginary axis, with the difference in degree of numerator and denominator no greater than one. But since one is even and the other odd they must in fact differ in degree by one. We are therefore in a position to write an expression for the driving point impedance of an LC network in a fairly explicit form.

$$Z(s) = H \frac{s(s^2+\omega_1^2)(s^2+\omega_3^2)\ldots}{(s^2+\omega_2^2)(s^2+\omega_4^2)\ldots} \qquad [10.4]$$

or

$$Z(s) = H \frac{(s^2+\omega_1^2)(s^2+\omega_3^2)\ldots}{s(s^2+\omega_2^2)(s^2+\omega_4^2)\ldots} \qquad [10.5]$$

where H is a scaling factor. The zeros occur at $s = \pm j\omega_1, \pm j\omega_3 \ldots$ and the poles at $s = \pm j\omega_2, \pm j\omega_4 \ldots$. The origin is either a pole or a zero and there is either a pole or a zero at infinity depending on whether the numerator or the denominator has the higher degree.

Now since $Z(s)$ must be positive real, and since its poles are on the imaginary axis they must be simple with real positive residues. We may therefore expand $Z(s)$ in partial fraction form as

$$Z(s) = k_\infty s + \frac{k_0}{s} + \frac{k_2}{s+j\omega_2} + \frac{k_2'}{s-j\omega_2} + \frac{k_4}{s+j\omega_4} + \frac{k_4'}{s-j\omega_4} + \ldots \quad [10.6]$$

where all the k are real and positive. Rewriting we find

$$Z(s) = k_\infty s + \frac{k_0}{s} + \frac{s(k_2+k_2')+j\omega_2(k_2'-k_2)}{s^2+\omega_2^2}$$
$$+ \frac{s(k_4+k_4')+j\omega_4(k_4'-k_4)}{s^2+\omega_4^2} + \ldots$$

But since $Z(s)$ is real for s real we require $k_2 = k_2'$, $k_4 = k_4' \ldots$ and so we may always write

$$Z(s) = k_\infty s + \frac{k_0}{s} + \frac{2k_2 s}{s^2+\omega_2^2} + \frac{2k_4 s}{s^2+\omega_4^2} + \ldots \qquad [10.7]$$

If there is no pole at infinity $k_\infty = 0$ and if there is no pole at the origin $k_0 = 0$.

Similarly, in general we may write

$$Y(s) = \alpha_\infty s + \frac{\alpha_0}{s} + \frac{2\alpha_1 s}{s^2 + \omega_1^2} + \frac{2\alpha_3 s}{s^2 + \omega_3^2} + \ldots \quad [10.8]$$

with all the α positive, since the poles are simple, $Y(s)$ being positive real. Again if $Y(s)$ has no pole at infinity α_∞ is zero and if no pole at the origin α_0 is zero.

The residues k and α may be calculated in a variety of ways. k_∞ and α_∞ are easily found by noting that

$$\lim_{s \to \infty} Z(s) = k_\infty s$$

$$\lim_{s \to \infty} Y(s) = a_\infty s \quad [10.9]$$

The residues k_0 and a_0 are given by

$$k_0 = sZ(s)\big|_{s=0}$$

$$\alpha_0 = sY(s)\big|_{s=0} \quad [10.10]$$

and the other residues by, for example,

$$k_2 = (s + j\omega_2)Z(s)\big|_{s=-j\omega_2} \quad [10.11]$$

Another expression for the residues is

$$k_i = \left[\frac{dY(s)}{ds}\right]^{-1}_{s=s_i} \quad [10.12]$$

where the pole (of Z) occurs at $s = s_i$ and k_i is the corresponding residue. Similarly

$$\alpha_j = \left[\frac{dZ(s)}{ds}\right]^{-1}_{s=s_j} \quad [10.13]$$

where the pole (of Y) occurs at $s = s_j$ and α_j is the corresponding residue.

Note that the residue of Z depends on the slope of Y at one of the zeros of Y (poles of Z) and vice versa.

10.1.1. SOME PROPERTIES OF LC IMPEDANCES

If $Z(s)$ is an LC impedance then $Z(j\omega)$ is purely imaginary and we may put $Z(j\omega) = jX(\omega)$

(i) $dX(\omega)/d\omega \geqq 0$.

From [10.7] we obtain

$$X(\omega) = k_\infty \omega - \frac{k_0}{\omega} + \sum_1^r \frac{2k_i \omega}{\omega_i^2 - \omega^2}$$

if these are r pairs of finite poles.

$$\frac{dX(\omega)}{d\omega} = k_\infty + \frac{k_0}{\omega^2} + \sum_1^r \frac{2k_i(\omega_i^2 + \omega^2)}{(\omega_i^2 - \omega^2)^2} \quad [10.14]$$

which is non-negative since all the k are positive. Note that it can only be zero at infinity.

(ii) $dX(\omega)/d\omega \geqq |X(\omega)|/\omega$.

From [10.13] and [10.14]

$$\frac{dX(\omega)}{d\omega} - \frac{X(\omega)}{\omega} = \frac{2k_0}{\omega^2} + \sum_1^r \frac{4k_i \omega^2}{(\omega_i - \omega^2)^2}$$

which is non-negative since the k are positive, and can only be zero at infinity,

$$\frac{dX(\omega)}{d\omega} + \frac{X(\omega)}{\omega} = 2k_\infty + \sum_1^r \frac{4k_i \omega^2}{(\omega_i^2 - \omega^2)^2}$$

which is again non-negative, and again can only be zero at infinity. Thus

$$\frac{dX(\omega)}{d\omega} \geqq \frac{|X(\omega)|}{\omega} \quad [10.15]$$

(iii) *If $s = \sigma$ then $Z(\sigma) \pm \sigma Z'(\sigma) \geqq 0$ where $Z'(\sigma) = dZ(\sigma)/d\sigma$.*

The proof follows the same lines as those given under (ii). Similar properties hold for $Y(s)$.

Since the slope of a reactance function is always positive on the imaginary axis it follows that the poles and zeros must interlace along this axis. Referring to Fig. 10.1(a) if points A and B are zeros of $Z(s)$ then since $Z(j\omega)$ is continuous there must be a turning point between A and B if there is no pole and so the slope would change sign. Therefore there must be a pole as shown in Fig. 10.1(b).

If points A and B are poles as shown in Fig. 10.2(a) and there is no zero between them then $X(\omega)$ being continuous must have the

TWO-ELEMENT-KIND SYNTHESIS 397

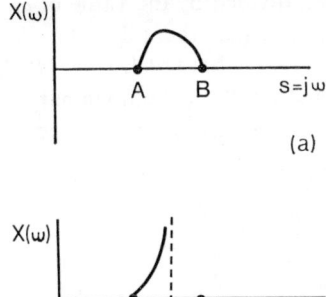

FIG. 10.1. (a) *A function with two successive zeros;* (b) *A function with two zeros separated by a pole.*

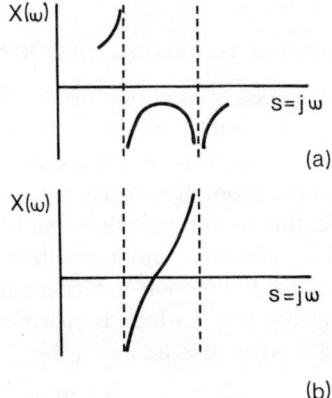

FIG. 10.2. (a) *A function with two successive poles;* (b) *A function with two poles separated by a zero.*

behaviour shown and so the slope changes sign. Thus there must be a zero between each pair of poles as shown in Fig. 10.2(*b*). Hence the poles and zeros are interlaced along the imaginary axis and $X(\omega)$ has the type of behaviour shown in Fig. 10.3. This figure is drawn from the case where the origin is a zero but the origin could equally well be a pole. In consequence of [*10.15*] the slope of the curve at any finite value of ω is greater than the value of the magnitude of the function at that point, divided by the value of ω. The same remarks regarding slope apply to $B(\omega)$ where $Y(j\omega) = jB(\omega)$.

It is interesting to observe that [10.12] and [10.13] give the residues in terms of slopes of Z or Y on the imaginary axis and this ties in with the fact that the slope has just been shown to be positive.

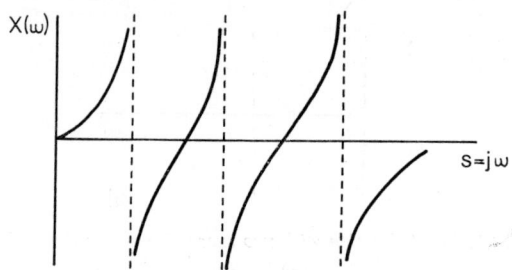

FIG. 10.3. *General form of a reactance function at real frequencies.*

10.1.2. REALISATION OF LC DRIVING POINT IMPEDANCES

We now know that a necessary condition on $Z(s)$ in order that it may represent the driving point impedance of an *LC* network is that it be a reactance function i.e. $Z(s)$ positive real and $Z(s)+Z(-s) \equiv 0$, and we have seen some of the properties of reactance functions. We shall now demonstrate that this condition is sufficient by finding a network composed of positive inductors and capacitors which is capable of realising any reactance function as its driving point impedance.

Let us further consider [10.7] which is reproduced below for the case of r pairs of finite poles of Z at $s = \pm j\omega_i$

$$Z(s) = k_\infty s + \frac{k_0}{s} + \sum_1^r \frac{2k_i s}{s^2+\omega_i^2}$$

Now if we could find a circuit whose impedance is $k_\infty s$, another whose impedance is k_0/s and a set of circuits whose impedance is $2k_i s/(s^2+\omega_i^2)$ and if these were connected in *series* then the driving point impedance of the resulting circuit would be $Z(s)$ and we would have achieved a realisation.

We can see by inspection that a circuit whose impedance is $k_\infty s$ is merely an inductor with $L = k_\infty$, and that a circuit whose impedance is k_0/s is a capacitor with $C = 1/k_0$. To find a circuit whose impedance is $2k_i s/(s^2+\omega_i^2)$ consider a parallel LC circuit as shown in Fig. 10.4.

$$Y = Cs + \frac{1}{Ls}$$

$$Z = \frac{Ls}{1+LCs^2}$$

$$= \frac{(1/C)s}{s^2+(1/LC)} \qquad [10.16]$$

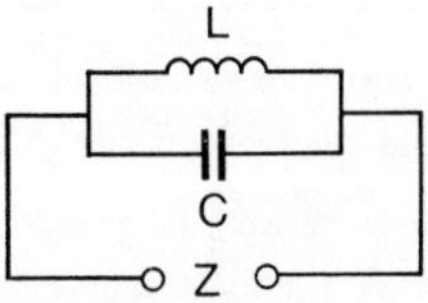

FIG. 10.4. *A parallel LC circuit.*

and if we put $1/C = 2k_i$ and $1/LC = \omega_i^2$ then the impedance of this circuit corresponds to $2k_i s/(s^2+\omega_i^2)$. Thus

$$C = \frac{1}{2k_i}$$

$$L = \frac{2k_i}{\omega_i^2} \qquad [10.17]$$

Note that in all cases the components are positive so long as k_∞, k_0 and k_i are positive which is assured if $Z(s)$ is a reactance function. Hence the network of Fig. 10.5 realises $Z(s)$ as its driving point

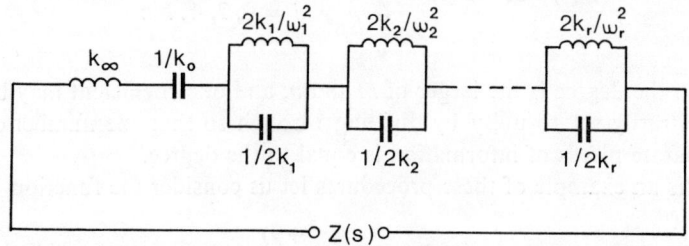

FIG. 10.5. *Realisation of a reactance function as a series circuit.*

impedance and so we have demonstrated the sufficiency of the reactance function condition for *LC* driving point impedances.

By an exactly dual process we could start from $Y(s)$ as given in [*10.8*], note that we require circuits whose *admittance* corresponds to the various terms and then connect such circuits in *parallel* to realise $Y(s)$, i.e. find an alternate realisation for $Z(s)$. A capacitor with $C = \alpha_\infty$ has an admittance $\alpha_\infty s$, an inductor with $L = 1/\alpha_0$ has an admittance α_0/s while a series *LC* circuit with $L = 1/2\alpha_j$ and $C = 2\alpha_j/\omega_j^2$ has an admittance $2\alpha_j s/(s^2+\omega_j^2)$ and the parallel connection of such circuits realises $Y(s)$ as shown in Fig. 10.6. The forms

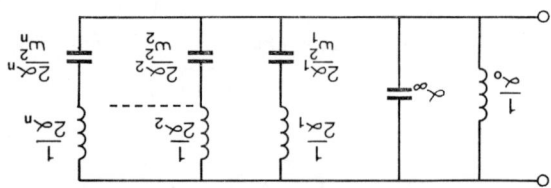

FIG. 10.6. *Realisation of a reactance function as a parallel circuit.*

shown in Figs. 10.5 and 10.6 are called *Foster forms*. These two forms are said to be *canonical* in that they are capable of realising *every* reactance function. They are said to be *minimal* in that they use the minimum possible number of circuit elements. This number is equal to the number of separate pieces of information in $Z(s)$ i.e. the sum of the number of residues and the number of finite pole pairs. Alternatively the number of separate pieces of information is equal to the degree of $Z(s)$. If we write

$$Z(s) \text{ or } Y(s) = \frac{\sum_{1}^{m} a_i s^i}{\sum_{0}^{n} b_j s^j} \quad \begin{array}{l} i = 1, 3, 5 \ldots m \\ j = 0, 2, 4 \ldots n \end{array}$$

then the degree is the larger of m and n, and one coefficient may be arbitrarily set to unity by dividing through so that the number of separate pieces of information is equal to the degree.

As an example of these procedures let us consider the function

$$Z(s) = \frac{s(s^2+9)}{(s^2+4)(s^2+16)} \qquad [10.18]$$

It is easily checked that $Z(s)$ is a reactance function. To realise $Z(s)$ in the form of Fig. 10.5 we expand $Z(s)$ about its poles. $Z(s)$ does not possess a pole at the origin or at infinity so we may write

$$Z(s) = \frac{2k_1 s}{s^2+4} + \frac{2k_2 s}{s^2+16}$$

$$k_1 = (s+j2)Z(s)\big|_{s=-j2}$$
$$= \tfrac{5}{24}$$
$$k_2 = (s+j4)Z(s)\big|_{s=-j4}$$
$$= \tfrac{7}{24}$$

Thus $Z(s)$ is realised by the series connection of two parallel LC circuits. From [10.17] one has $C = 12/5$ Farads, $L = 5/48$ Henrys, and the other has $C = 12/7$ Farads, $L = 7/192$ Henrys, as shown in Fig. 10.7.

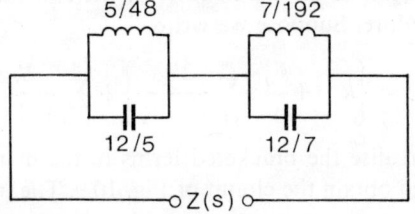

FIG. 10.7. *Realisation of the impedance function of* [10.18] *in series form.*

For the alternative realisation we expand $Y(s) = 1/Z(s)$ about its poles as follows

$$Y(s) = \alpha_\infty s + \frac{\alpha_0}{s} + \frac{2\alpha_1 s}{s^2+9} \qquad [10.19]$$

as $s \to \infty$ $Y(s) \to s$ so $\alpha_\infty = 1$

$$\alpha_0 = sY(s)\big|_{s=0}$$
$$= \tfrac{64}{9}$$
$$\alpha_1 = (s+j3)Y(s)\big|_{s=-j3}$$
$$= \tfrac{35}{18}$$

Thus

$$Y(s) = s + \frac{64/9}{s} + \frac{(35/9)s}{s^2+9} \qquad [10.20]$$

Now a capacitor of 1 Farad has an admittance s, an inductor with $L = 9/64$ has an admittance $64/9s$ while, from Fig. 10.6 the last term is produced by the admittance of a series LC circuit with $L = 9/35$ Henrys and $C = 35/81$ Farads. The final circuit which provides an alternative realisation of $Z(s)$ is shown in Fig. 10.8.

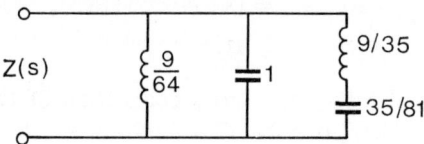

FIG. 10.8. *Realisation of the impedance function of* [10.18] *in parallel form.*

It is also possible to have a mixture of the two Foster forms in a synthesis procedure. Suppose we write

$$Z(s) = \left\{ k_\infty s + \frac{k_0}{s} + \sum_1^m \frac{2k_i s}{s^2+\omega_i^2} \right\} + \sum_{m+1}^n \frac{2k_i s}{s^2+\omega_i^2}$$

Now we may realise the bracketed terms in the manner of the first Foster form and obtain the circuit of Fig. 10.9. The impedance $Z_1(s)$

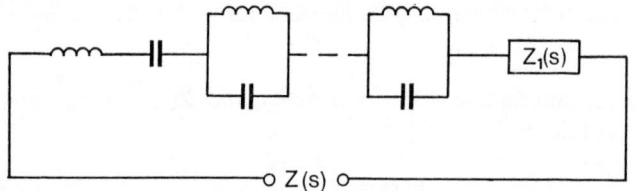

FIG. 10.9. *An illustration of a mixed series/parallel realisation of a reactance function.*

represents those terms outside the bracket. Now we may expand $Y_1(s) = 1/Z_1(s)$ about its poles and realise it in the second Foster form and so obtain a realisation of the original $Z(s)$ in mixed form. We may, of course, switch from one form to the other in as many ways as we like and thus produce a great variety of networks to realise any given reactance function.

TWO-ELEMENT-KIND SYNTHESIS

It is instructive to look at these realisation procedures in a different way. If

$$Z(s) = k_\infty s + \frac{k_0}{s} + \sum_1^r \frac{2k_i s}{s^2 + \omega_i^2} \qquad [10.21]$$

then

$$Z_1(s) = Z(s) - ks \qquad [10.22]$$

is a reactance function if $k \leqq k_\infty$ since it is of the same form as $Z(s)$ except that the residue at infinity may possibly be zero. Similarly $Z_1(s) = Z(s) - k/s$ is a reactance function if $k \leqq k_0$ and $Z_1(s) = Z(s) - 2ks/(s^2 + \omega_j^2)$ is a reactance function if ω_j is one of the poles and $k \leqq k_j$. This process of forming $Z_1(s)$ is called *extraction* of a pole from $Z(s)$. If k is equal to the appropriate residue in $Z(s)$ it is called complete extraction and if k is smaller than this it is called partial extraction. A similar procedure can be applied to $Y(s)$. Now we may say that the first Foster form is generated as follows: the pole at infinity (if any) is completely extracted from $Z(s)$ to leave a remainder $Z_1(s)$ (a reactance function). The pole at the origin (if any) is completely extracted from $Z_1(s)$ to leave a remainder $Z_2(s)$ (again a reactance function). One of the finite poles is completely extracted from $Z_2(s)$ leaving a remainder $Z_3(s)$ (a reactance function) from which a further finite pole is extracted and so on until the remainder is zero. This is illustrated in Fig. 10.10, and is a rather general type of approach which we encounter frequently. Something recognisable is extracted from the given function leaving a remainder which satisfies the same necessary conditions as the original but which is somehow simpler (or at least no more complicated) and amenable to further treatment. Repetition of this process should eventually lead to a remainder which is itself recognisable and so complete the realisation procedure. It is worth noting that at any stage the remainder may be treated in any legitimate manner and one does not have to use the same extraction procedure at every stage. For example, we might start with $Z(s)$ and completely remove poles of impedance for several cycles ending with a remainder $Z_r(s)$. We could then take $Y_r(s) = 1/Z_r(s)$ and completely remove poles of admittance for several cycles leaving a remainder $Y_n(s)$. We could then take $Z_n(s) = 1/Y_n(s)$ and revert to removing poles of impedance, and so on. We can therefore see that there are many possible ways of

E

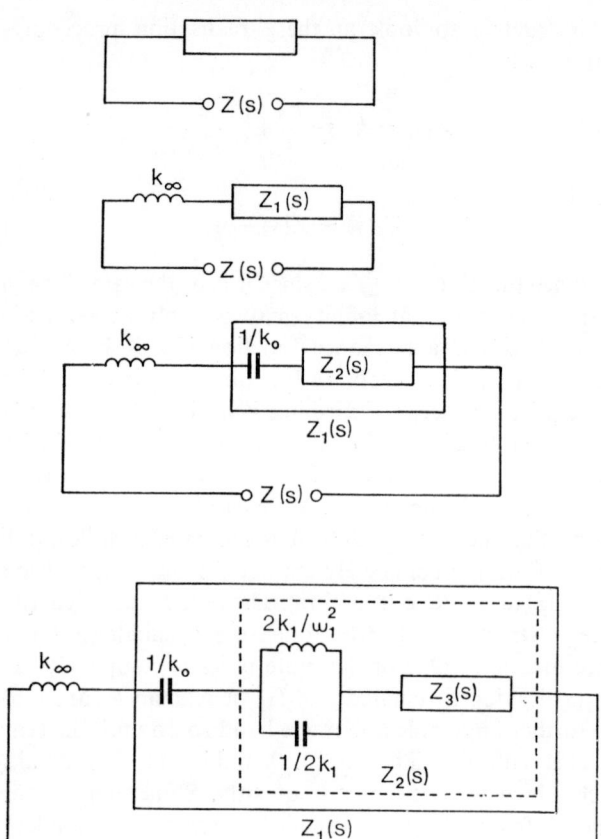

FIG. 10.10. *Illustration of the concept of extraction of poles from an impedance function.*

realising a given function, and indeed if we use partial pole removal there are an infinite number of possible circuits corresponding to a given driving point impedance.

10.1.3. THE CAUER FORMS

Another form of synthesis for *LC* driving point impedances is the Cauer or ladder form. These are two possibilities and both are canonic and minimal. One is based on extraction of poles at the origin and the other on extraction of poles at infinity. We consider first the *extraction of poles at infinity*. If $Z(s)$ is a reactance function then either $Z(s)$ or $Y(s)$ has a pole at infinity. Assume $Z(s)$ has this pole. We then write

$$Z(s) = k_\infty s + Z_1(s)$$

where k_∞ is the residue of $Z(s)$ at the pole. In network terms this is shown in Fig. 10.11(a). Now $Z_1(s)$ clearly does not have a pole at infinity but since it is a reactance function it must therefore have a zero at infinity and so $Y_1(s) = 1/Z_1(s)$ has a pole at infinity. We then write

$$Y_1(s) = \alpha_\infty s + Y_2(s)$$

where α_∞ is the residue of Y_1 at infinity. This leads to the network of Fig. 10.11(b). Now $Y_2(s)$ does not have a pole at infinity, therefore,

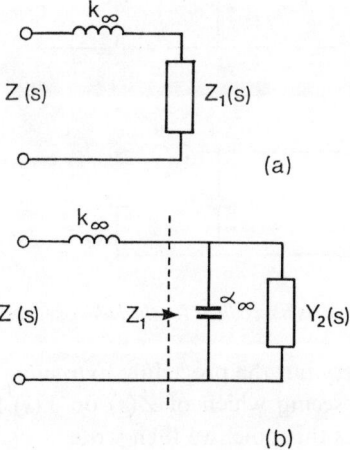

FIG. 10.11. *Extraction of poles at infinity* (a) *from an impedance and* (b) *from an admittance.*

being a reactance function it must have a zero at infinity so that $Z_2(s) = 1/Y_2(s)$ has a pole at infinity. We then write

$$Z_2(s) = k'_\infty s + Z_3(s)$$

and the cycle repeats. Thus we obviously generate the network of Fig. 10.12. Now, if originally it were $Y(s)$ rather than $Z(s)$ which had the pole at infinity, the network would commence with a shunt capacitor rather than a series inductor. Each step in the procedure reduces the degree of the remaining function by one and the process terminates when a remainder $Z_r = 0$ or a remainder $Y_r = 0$. This leads to either the ending shown in Fig. 10.13(a) (when $Z_r = 0$) or that shown in Fig. 10.13(b) (where $Y_r = 0$).

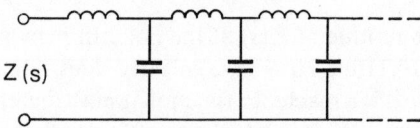

Fig. 10.12. *Ladder network resulting from synthesis by extraction of poles at infinity.*

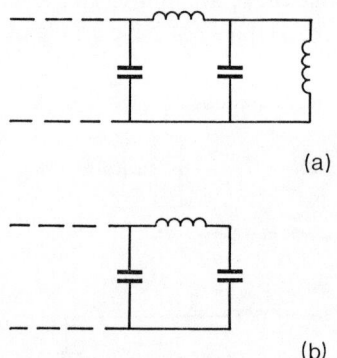

Fig. 10.13. *The possible endings for the ladder network of Fig. 10.12.*

If we wish to carry out the procedure *extracting poles at the origin* then we begin by seeing which of $Z(s)$ or $Y(s)$ has a pole at the origin. Say $Z(s)$ has this pole, we then write

$$Z(s) = (k_0/s) + Z_1(s)$$

where k_0 is the residue at this pole. $Z_1(s)$ now has no pole at the origin so it must have a zero there and so $Y_1(s)$ has a pole. We then write

$$Y_1(s) = (\alpha_0/s) + Y_2(s)$$

where α_0 is the residue at the pole. Now $Y_2(s)$ has no pole at the origin so $Z_2(s)$ must have such a pole and we repeat the cycle starting with $Z_2(s)$. This procedure leads to the realisation of Fig. 10.14. Again if $Y(s)$ had the pole at the origin originally rather than $Z(s)$ the

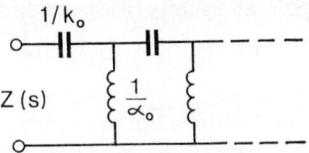

FIG. 10.14. *Ladder network resulting from synthesis by extraction of poles at the origin.*

first series capacitor would be missing. The procedure ends when a remainder $Z_r(s)$ or $Y_r(s)$ is zero, to give respectively the endings shown in Fig. 10.15(a) and (b).

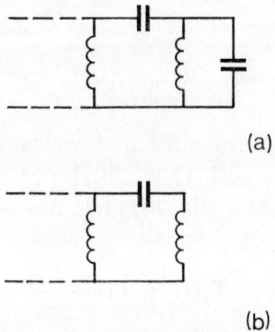

FIG. 10.15. *The possible endings for the ladder network of Fig. 10.14.*

To illustrate these procedures we take the example of [*10.18*]. Now suppose we wish to extract poles at infinity then we see that it is $Y(s)$ which possesses this pole and the residue is 1. Thus

$$Y(s) = s + Y_1(s)$$

and

$$Y_1(s) = \frac{11s^2 + 64}{s^3 + 9s}$$

$Z_1(s)$ has a pole at infinity with residue $1/11$ so we write

$$Z_2(s) = Z_1(s) - s/11$$
$$= \frac{35s/11}{11s^2 + 64}$$

Now $Y_2(s)$ has a pole at infinity with residue 121/35 so we write
$$Y_3(s) = Y_2(s) - 121s/35$$
$$= \frac{704}{35s}$$

$Z_3(s)$ has a pole at infinity with residue 35/704 and so we write
$$Z_4(s) = Z_3(s) - 35s/704$$
$$= 0$$

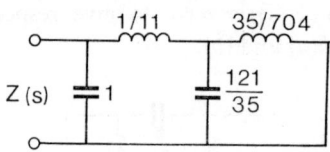

FIG. 10.16. *Realisation of the impedance given in [10.18] by extraction of poles at infinity.*

Thus the procedure terminates and yields the circuit shown in Fig. 10.16. Now if we wish to synthesise by extraction of poles at the origin we note that again $Y(s)$ has this pole with residue 64/9. We then write
$$Y_1(s) = Y(s) - \frac{64}{9s}$$
$$= \frac{s^3 + (116/9)s}{s^2 + 9}$$

Now $Z_1(s)$ has a pole at the origin with residue 81/116 and so we write
$$Z_2(s) = Z_1(s) - \frac{81}{116s}$$
$$= \frac{(35/116)s}{s^2 + (116/9)}$$

Now $Y_2(s)$ has a pole at the origin with residue $(116)^2/315$ so we write
$$Y_3(s) = Y_2(s) - \frac{(116)^2}{315s}$$
$$= \frac{116s}{35}$$

Now $Z_3(s)$ has a pole at the origin with residue 35/116 and so we write

$$Z_4(s) = Z_3(s) - \frac{35}{116s}$$
$$= 0$$

Thus we obtain the circuit of Fig. 10.17.

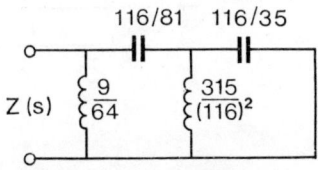

FIG. 10.17. *Realisation of the impedance given in [10.18] by extraction of poles at the origin.*

The procedures for finding the Cauer forms may be written in a somewhat more compact manner as follows. What in fact we are doing is writing, say, $Z(s)$ in the form

$$Z(s) = k_1 s + Z_1(s)$$
$$= k_1 s + \frac{1}{Y_1(s)}$$
$$= k_1 s + \frac{1}{k_2 s + Y_2(s)}$$
$$= k_1 s + \frac{1}{k_2 s + [1/Z_2(s)]}$$
$$= k_1 s + \frac{1}{k_2 s + [1/\{k_3 s + Z_3(s)\}]}$$

and so on, in the case where we extract poles at infinity. The complete realisation therefore is obtained by expressing $Z(s)$ as

$$Z(s) = k_1 s \cfrac{1}{k_2 s + \cfrac{1}{k_3 s + \cfrac{1}{k_4 s}}}$$

.

. [10.23]

This is known as a continued fraction expansion of $Z(s)$. The coefficients k are the values of the elements in the circuit.

In the case where we extract poles at the origin a similar argument shows that we are in fact writing

$$Z(s) = \frac{\alpha_1}{s} + \cfrac{1}{\cfrac{\alpha_2}{s} + \cfrac{1}{\cfrac{\alpha_3}{s} + \cfrac{1}{\cfrac{\alpha_4}{s} + \cdots}}}$$

[10.24]

and the coefficients α are the reciprocals of the values of the circuit elements. If $Z(s)$ is a reactance function then the coefficients k and α are, of course, positive.

There is a more systematic way to calculate the coefficients than that previously described. The method is best explained by an example. Suppose we wish to extract poles at infinity and the given impedance is

$$Z(s) = \frac{(s^2+2)(s^2+4)}{s(s^2+3)}$$

which has a pole at infinity. We begin by writing numerator and denominator in descending powers of s to give

$$Z(s) = \frac{s^4 + 6s^2 + 8}{s^3 + 3s}$$

Now the first step is to write $Z(s)$ as $k_1 s + Z_1(s)$ and k_1 is the ratio of the coefficients of the highest powers of s in numerator and denominator. We therefore divide the denominator into the numerator

$$\begin{array}{r} s \\ s^3+3s \overline{\smash{\big)}\, s^4+6s^2+8} \\ \underline{s^4+3s^2} \\ 3s^2+8 \end{array}$$

then
$$Z(s) = s + \frac{3s^2+8}{s^3+3s}$$
$$= s + Z_1(s)$$

We must now extract the pole from $Y_1(s)$ which we may accomplish in the same way

$$Y_1(s) = \frac{s^3+3s}{3s^2+8}$$

and we divide the denominator into the numerator as before

$$3s^2+8 \overline{\smash{\big)}\, s^3+3s} \quad \text{quotient } s/3$$
$$s^3 + \frac{8s}{3}$$
$$\frac{s}{3}$$

so
$$Y_1(s) = \frac{s}{3} + \frac{s/3}{3s^2+8}$$
$$= \frac{s}{3} + Y_2(s)$$

We must now extract the pole from $Z_2(s)$ and we proceed as before by dividing the denominator into the numerator

$$\frac{s}{3} \overline{\smash{\big)}\, 3s^2+8} \quad \text{quotient } 9s$$
$$3s^2$$
$$8$$

so
$$Z_2(s) = 9s + \frac{24}{s}$$
$$= 9s + Z_3(s)$$

and we remove the pole at infinity from $Y_3(s)$ to leave a zero remainder.

The whole process may be described as follows: write numerator and denominator in descending powers of s. Divide the denominator into the numerator to find the quotient $k_1 s$. Divide the remainder into the previous divisor to find the second quotient $k_2 s$ and a new remainder. Divide that remainder into the immediately preceding divisor and continue the process until a zero remainder is obtained. The quotients are the element values of the network and are alternately inductance and capacitance. If the original function is an impedance the first quotient is an inductance and if the original function is an admittance the first quotient is a capacitance. Thus for the example just considered the process gives

$$
\begin{array}{r}
s \\
s^3+3s \overline{\big) s^4+6s^2+8} \\
\underline{s^4+3s^2} \quad \big| s/3 \\
3s^2+8 \overline{\big) s^3+3s} \\
\underline{s^3+\dfrac{8s}{3}} \,\bigg|\, 9s \\
\dfrac{s}{3} \overline{\big) 3s^2+8} \\
3s^2 \quad \big| s/24 \\
8 \,\bigg/\, \dfrac{s}{3} \\
\dfrac{s}{3} \\
\overline{0}
\end{array}
$$

and since we started with Z the first coefficient is a series inductor with $L = 1$ leading to the circuit of Fig. 10.18.

If we wish to extract poles at the origin then we write numerator and denominator in ascending powers of s and starting with Z or Y

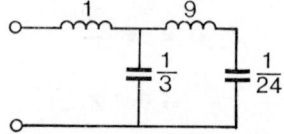

Fig. 10.18. *An illustration of synthesis by continued fraction expansion according to poles at infinity.*

(whichever has the pole at the origin) we divide the denominator into the numerator and proceed as before. For the example just considered Z has the pole at the origin and

$$Z = \frac{8+6s^2+s^4}{3s+s^3}$$

or

$$Z = 3s+s^3 \overline{\smash{\big)}\,8+6s^2+s^4} \frac{8/3s}{}$$

$$\begin{array}{c|c} 8+\dfrac{8}{3}s^2 & \dfrac{9}{10s} \\ \hline \dfrac{10}{3}s^2+s^4 \Big/ 3s+s^3 \\ 3s+\dfrac{9}{10}s^3 & \dfrac{100}{3s} \\ \hline \dfrac{1}{10}s^3 \Big/ \dfrac{10}{3}s^2+s^4 \\ \dfrac{10}{3}s^2 & \dfrac{1}{10s} \\ \hline s^4 \Big/ \dfrac{1}{10}s^3 \\ \dfrac{1}{10}s^3 \\ \hline 0 \end{array}$$

The first element is a series capacitor $C = 3/8$ and the elements are alternately capacitors and inductors. The resulting circuit is shown in Fig. 10.19.

The process of division-inversion-division as just described generates the coefficients of the continued fraction expansion of a function. When $Z(s)$ is a reactance function the coefficients are

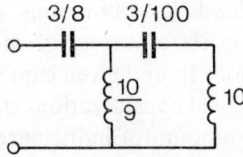

Fig. 10.19. *A realisation of the impedance of the circuit of Fig. 10.18 by continued fraction expansion according to poles at the origin.*

positive and indeed this type of expansion provides a very convenient means for checking whether a function is in fact a reactance function without the necessity of determining the poles and zeros which may be difficult if the degree is high. The method can also be used to check whether a polynomial is Hurwitz since, if it is, then the ratio of its even to its odd part must be a reactance function and we have merely to carry out a continued fraction expansion of this and check whether the coefficients are positive.

Another useful form of realisation is the following. We write

$$Z(s) = k_\infty s + \frac{k_0}{s} + Z_1(s)$$

where k_∞ or k_0 may be zero. In any case $Z_1(s)$ has no pole at the origin or at infinity and is of course a reactance function. Hence $Y_1(s)$ must have a pole at the origin and at infinity so we write

$$Y_1(s) = \alpha_\infty s + \frac{\alpha_0}{s} + Y_2(s)$$

Now $Y_2(s)$ has no pole at the origin or at infinity so $Z_2(s)$ must have both. We then repeat the cycle starting with $Z_2(s)$. The resulting circuit is shown in Fig. 10.20. The first series inductor or capacitor may be absent.

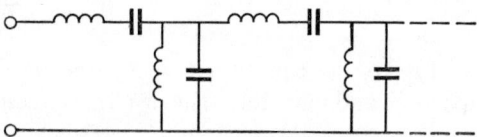

Fig. 10.20. *Realisation of a reactance function by extraction of poles at the origin and infinity.*

We have now seen several ways in which LC driving point impedances may be realised. Combinations of these methods are possible so that there exists a variety of minimal realisations for a given reactance function. In any given case the choice of a method depends on various practical considerations such as for example how to obtain the minimum amount of inductance, how to minimise the ratio of the largest to the smallest component value, or how to actually construct the circuit.

10.2. LC Transfer Functions

So far we have been dealing exclusively with driving point impedances. We now turn our attention to LC (lossless) two-port networks. Such networks are in themselves interesting and are of great importance since they form the basis upon which the synthesis of RLC driving point impedances due to Darlington (to be discussed in Chapter 11) and the synthesis of general transfer functions (to be discussed in Chapter 12) are based.

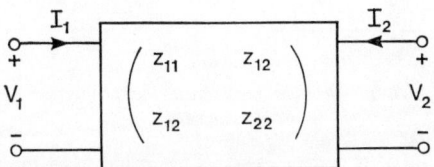

FIG. 10.21. *A lossless two-port defined by its z-matrix.*

We consider a lossless two port defined, for example, by its z-matrix as shown in Fig. 10.21. Now we know immediately that z_{11} and z_{22} must be reactance functions since they are the (open-circuit) driving point impedances of an LC network. However, we have no knowledge of z_{12}, since it is a transfer impedance, or of the necessary relationships between z_{11}, z_{22} and z_{12}. To investigate these questions we consider the circuit of Fig. 10.22. Now

$$V = V_A + V_B$$

and

$$V_A/n_1 = z_{11}(n_1 I) + z_{12}(n_2 I)$$
$$V_B/n_2 = z_{12}(n_1 I) + z_{22}(n_2 I)$$

thus

$$V = V_A + V_B = I(n_1^2 z_{11} + 2 n_1 n_2 z_{12} + n_2^2 z_{22})$$

and

$$Z = V/I = n_1^2 z_{11} + 2 n_1 n_2 z_{12} + n_2^2 z_{22} \qquad [10.25]$$

Now Z is the driving point impedance of a lossless network and so it must be a reactance function. In fact Z must be a reactance function for any arbitrary real choice of the turns ratios of the transformers.

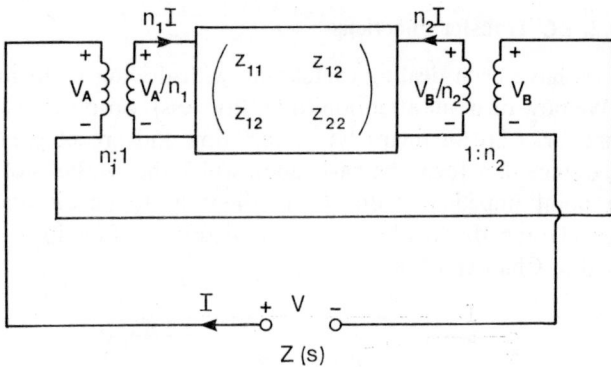

Fig. 10.22. *Circuit used to investigate the properties of z_{12} of a lossless network.*

Now let z_{12} have a pole $s = s_1$. This must also be a pole of Z (if z_{12} is infinite so is Z, at least for some choice of n_1 and n_2). But Z is a reactance function for all n_1 and n_2, hence the poles of z_{12} must lie on the imaginary axis and must be simple. The same is not true of the zeros of z_{12} since zeros of z_{12} are not zeros of Z. Now if z_{12} has a pole at $s = \pm j\omega_i$ with residue k_{12} and z_{11} and z_{22} do not have this pole then the residue of $Z(s)$ at this pole is $2n_1 n_2 k_{12}$. Whether k_{12} is positive or negative we can then always choose values for n_1 and n_2 such that $2n_1 n_2 k_{12}$ is negative and so $Z(s)$ would have a negative residue which is not in accordance with the fact that $Z(s)$ is a reactance function. Thus z_{12} cannot have a pole which is not a pole of z_{11} and z_{22}. Suppose this pole is possessed by z_{11} and not z_{22} and the residue of z_{11} at this pole is k_{11} (which must be positive since z_{11} is a reactance function). Then the residue of $Z(s)$ at this pole is $n_1^2 k_{11} + 2n_1 n_2 k_{12}$. Again we can always choose n_1 and n_2 so that this is negative and so we conclude that all poles of z_{12} must be simultaneously poles of both z_{11} and z_{22}. It is, of course, possible for z_{11} or z_{22} to possess poles not common with the other two. If, for example, z_{11} has a pole with residue k_{11} ($k_{11} > 0$) which is not a pole of z_{12} or z_{22} then the residue of $Z(s)$ at this pole is $n_1^2 k_{11}$ which is positive for all n_1 such poles are called 'private poles' (of z_{11} or z_{22}).

We now consider the general case of a pole at $s = \pm j\omega_1$ common to z_{11}, z_{12} and z_{22} and we let the three residues at this pole be k_{11}, k_{12} and k_{22} respectively. The residue of $Z(s)$ is then

$$k = n_1^2 k_{11} + 2n_1 n_2 k_{12} + n_2^2 k_{22} \qquad [10.26]$$

TWO-ELEMENT-KIND SYNTHESIS

and we require $k \geqq 0$ for all real choices of n_1 and n_2. This is similar to the problem encountered in Chapter 2 of Vol. 1, where the energy stored in coupled coils was shown to be $\varepsilon = L_1 i_1^2 + 2M i_1 i_2 + L_2 i_2^2$ and we required $\varepsilon \geqq 0$ for all real currents i_1 and i_2. In the present case we adopt the same approach. Now if we require that k never be negative it will be sufficient if it may never be zero for any *real* choice of n_1 and n_2. We therefore put $k = 0$ and solve for n_1 to give

$$n_1 = \frac{n_2}{k_{11}} \{-k_{12} \pm \sqrt{(k_{12}^2 - k_{11}k_{22})}\} \qquad [10.27]$$

Now there can be no pair of *real* values of n_1 and n_2 which satisfy this if

$$k_{11}k_{22} - k_{12}^2 > 0 \qquad [10.28]$$

since then the quantity in brackets is complex. So if [10.28] is satisfied k can never be zero and then can certainly never be negative. We may, in fact, relax condition [10.28] slightly since if we allow the equality sign [10.26] may be written as $(n_1\sqrt{k_{11}} + n_2\sqrt{k_{22}})^2$ which is clearly non-negative. Therefore the strictly necessary condition is

$$k_{11}k_{22} - k_{12}^2 \geqq 0 \qquad [10.29]$$

at every pole. When $k_{11}k_{22} = k_{12}^2$ at every pole the network is said to be compact. Note that k_{12} may be either positive or negative since only k_{12}^2 is involved in [10.29]. If at a particular pole $k_{11}k_{22} > k_{12}^2$ then we may always partially extract the pole from z_{11} or z_{22} so as to leave a remaining network which is compact. For example we may remove a term $2k'_{11}s/(s^2 + \omega_1^2)$ from z_{11} (or z_{22}) such that

$$(k_{11} - k'_{11})k_{22} = k_{12}^2$$

This is illustrated in Fig. 10.23.

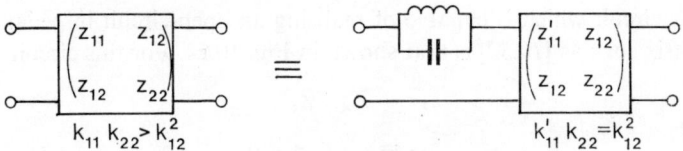

FIG. 10.23. *Partial removal of a pole to create a compact network.*

10.2.1. SUFFICIENCY OF THE NECESSARY CONDITIONS

We must now prove the sufficiency of the condition [10.29]. To do this we must find a network capable of realising any (z) matrix satisfying the necessary conditions. In general we may write

$$z_{11} = k_{11}^{(\infty)}s + \frac{k_{11}^{(o)}}{s} + \sum_{1}^{n} \frac{2k_{11}^{(i)}s}{s^2 + \omega_i^2}$$

$$z_{22} = k_{22}^{(\infty)}s + \frac{k_{22}^{(o)}}{s} + \sum_{1}^{n} \frac{2k_{22}^{(i)}s}{s^2 + \omega_i^2}$$

$$z_{12} = k_{12}^{(\infty)}s + \frac{k_{12}^{(o)}}{s} + \sum_{1}^{n} \frac{2k_{12}^{(i)}s}{s^2 + \omega_i^2}$$

[10.30]

with $k_{11}k_{22} - k_{12}^2 \geq 0$ at every pole. If z_{11} or z_{22} have any private poles they are first removed as series elements consisting of parallel LC circuits in the usual way. The set of equations [10.30] may be rewritten in matrix form as

$$(z) = s\begin{pmatrix} k_{11}^{(\infty)} & k_{12}^{(\infty)} \\ k_{12}^{(\infty)} & k_{22}^{(\infty)} \end{pmatrix} + \frac{1}{s}\begin{pmatrix} k_{11}^{(o)} & k_{12}^{(o)} \\ k_{12}^{(o)} & k_{22}^{(o)} \end{pmatrix} + \sum_{1}^{n} \frac{2s}{s^2 + \omega_i^2}\begin{pmatrix} k_{11}^{(i)} & k_{12}^{(i)} \\ k_{12}^{(i)} & k_{22}^{(i)} \end{pmatrix}$$

[10.31]

To realise (z) we need only to find a two-port network to realise each term in the matrix expansion [10.31] and then connect these two-ports in series as explained in Chapter 5. The various open circuit impedance matrices which must be realised may be written as

$$f(s)\begin{pmatrix} k_{11} & k_{12} \\ k_{12} & k_{22} \end{pmatrix}$$

[10.32]

where $f(s) = s$, $1/s$, or $2s/(s^2 + \omega_i^2)$ and $k_{11}k_{22} - k_{12}^2 \geq 0$, $k_{11} > 0$, $k_{22} > 0$.

A circuit which is capable of realising an open-circuit impedance matrix such as [10.32] is that shown in Fig. 10.24. For this circuit

$$z_{11} = Z_A + Z_C$$
$$z_{22} = n^2(Z_B + Z_C)$$
$$z_{12} = nZ_C$$

[10.33]

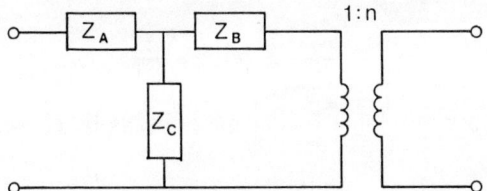

FIG. 10.24. *A circuit capable of realising the open-circuit impedance matrix of [10.32].*

We now write $Z_A = af(s)$, $Z_B = bf(s)$, $Z_C = cf(s)$, so that

$$z_{11} = (a+c)f(s) = k_{11}f(s)$$
$$z_{22} = n^2(b+c)f(s) = k_{22}f(s)$$
$$z_{12} = ncf(s) = k_{12}f(s)$$

Thus we must solve

$$a+c = k_{11}$$
$$n^2(b+c) = k_{22}$$
$$nc = k_{12} \qquad [10.34]$$

Clearly since we require a, b, and c to be positive in order that Z_A, Z_B, Z_C be realisable, n must have the same sign as k_{12}. Thus

$$c = k_{12}/n = \frac{|k_{12}|}{|n|}$$

$$b = \frac{k_{22}}{n^2} - \frac{k_{12}}{n} = \frac{k_{22}}{|n|^2} - \frac{|k_{12}|}{|n|}$$

$$a = k_{11} - \frac{k_{12}}{n} = k_{11} - \frac{|k_{12}|}{|n|} \qquad [10.35]$$

Thus we require

$$k_{11} \geqq \frac{|k_{12}|}{|n|}$$

$$k_{22} \geqq |n|\,|k_{12}|$$

or

$$\frac{|k_{12}|}{k_{11}} \leq |n| \leq \frac{k_{22}}{|k_{12}|} \qquad [10.36]$$

so that a (positive) value for $|n|$ may always be chosen to satisfy [*10.36*] if

$$\frac{|k_{12}|}{k_{11}} \leq \frac{k_{22}}{|k_{12}|}$$

or

$$k_{11}k_{22} \geq |k_{12}|^2$$

i.e.

$$k_{11}k_{22} - k_{12}^2 \geq 0$$

with $k_{11}, k_{22} > 0$.

a, b and c are then positive and Z_A, Z_B and Z_C are realisable as inductors if $f(s) = s$, as capacitors if $f(s) = 1/s$, or as parallel LC circuits if $f(s) = 2s/(s^2 + \omega_i^2)$. The individual two-ports may then be connected in series to realise the given z-matrix thus demonstrating the sufficiency of the conditions.

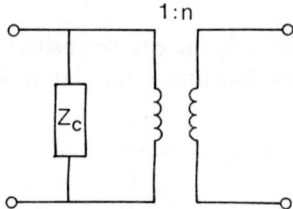

FIG. 10.25. *Realisation of a compact open-circuit impedance matrix.*

In practice we may always choose the value of $|n|$ in order to simply the circuit of Fig. 10.24. If $|n| = k_{22}/|k_{12}|$ then $b = 0$ and Z_B disappears, while if $|n| = |k_{12}|/k_{11}$ then $a = 0$ and Z_A disappears. If the network is compact then since $k_{11}k_{22} = k_{12}^2$ the limits in [*10.36*] coincide and $|n| = k_{22}/|k_{12}| = |k_{12}|/k_{11}$ so that $a = b = 0$ and the circuit of Fig. 10.25 results. If k_{12} is positive and if $k_{11} > k_{12}$ and $k_{22} > k_{12}$ then n is positive and the range in [*10.36*] includes the value unity, so that the ideal transformer has $n = 1$ and may be removed.

By a process dual to the above we may also show the same conditions to be necessary and sufficient for the *y*-matrix. The resulting general network is shown in Fig. 10.26, and these networks are connected in parallel. Again we may always choose the value of n such that Y_A or Y_B is zero, and if the network is compact both Y_A and Y_B are zero.

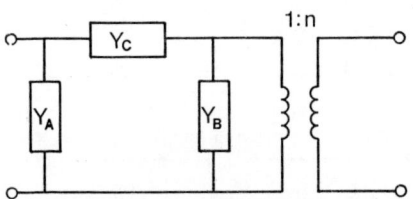

FIG. 10.26. *A circuit capable of realising a general short-circuit admittance matrix.*

As an example of this procedure let us realise the impedance matrix

$$(z) = s \begin{pmatrix} 1 & 2 \\ 2 & 4 \end{pmatrix} + \frac{1}{s}\begin{pmatrix} 1 & 1 \\ 1 & 1 \end{pmatrix} \qquad [10.37]$$
$$\quad\;\; \text{I} \qquad\qquad \text{II}$$

The resulting network will be compact since $k_{11}k_{22} - k_{12}^2 = 0$ at each pole. Considering first matrix I we have the circuit of Fig. 10.27(*a*). Since k_{12} is positive n is positive and from [*10.36*] we have

$$n = \frac{k_{12}}{k_{11}} = 2 \qquad [10.38]$$

and from [*10.35*]

$$L = k_{12}/n = 1 \qquad [10.39]$$

For matrix II we have the circuit of Fig. 10.27(*b*) and from [*10.36*] $n = 1$ while [*10.35*] gives

$$C = k_{12}/n = 1 \qquad [10.40]$$

The two circuits are then connected in series as shown in Fig. 10.27(*c*) and the 1:1 transformer associated with the capacitor has been removed. This is then a realisation of (z) as given in [*10.37*].

In the case of a circuit such as Fig. 10.27(*a*) the ideal transformer may always be removed by means of the equivalence with a pair of

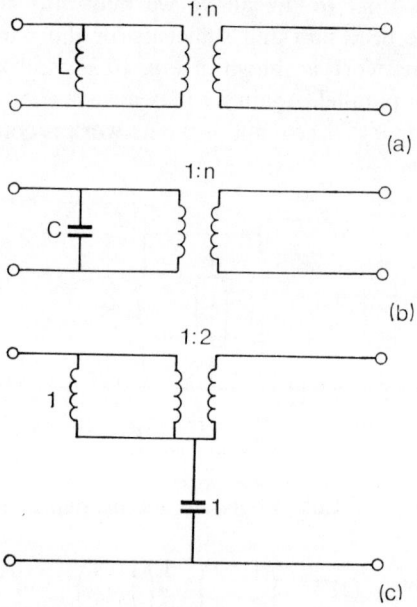

FIG. 10.27. *Realisation of the open-circuit impedance matrix of* [*10.37*] *(a) matrix I (b) matrix II (c) the complete matrix.*

coupled coils shown in Fig. 10.28. Comparison of the two circuits gives

$$z_{11} = Ls = L_p s$$
$$z_{22} = n^2 Ls = L_s s$$
$$z_{12} = nLs = Ms \qquad [10.41]$$

Hence

$$L_p = L, \qquad L_s = n^2 L, \qquad M = nL \qquad [10.42]$$

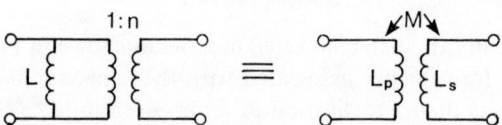

FIG. 10.28. *Equivalence between a shunt inductor with ideal transformer and a pair of coupled coils.*

so that the coils are perfectly coupled ($L_pL_s = M^2$). Using this equivalence the circuit of Fig. 10.27(c) becomes that shown in Fig. 10.29. This is called a *Brune Section* ($M > 0$). If M is negative this circuit is called the *Darlington C-section* (see the examples at the end of this chapter). Both of these circuits have general importance in the synthesis of *RLC* networks, as we shall see later.

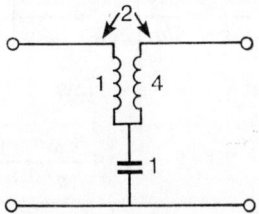

FIG. 10.29. *A circuit with coupled coils equivalent to that in Fig. 10.27(c).*

In the general procedure presented here it should be noted that the ideal transformer plays an essential part and cannot in general be dispensed with, except that in special cases it turns out to be 1:1 and in combination with an inductor may be replaced by a pair of perfectly coupled coils.

10.3. *RC* Driving Point Impedances

This class of networks is of considerable practical as well as theoretical importance. From a practical viewpoint the difficulty of constructing inductors without loss means that the realisation of networks involving inductors is seldom very exact, while resistors and capacitors can be made with a high degree of 'purity'. As we shall see later the properties of *RC* networks can be derived from those of corresponding *LC* networks by a transformation but they are of sufficient importance to be studied in their own right.

Now the characteristic of an *RC* network is that it has no inductors so that the stored magnetic energy is zero. Thus from [*9.34*] $T_0 \equiv 0$ and any driving point impedance may be written as

$$Z = F_0 + (V_0/s) \qquad [10.43]$$

The zeros are therefore given by

$$s = -(V_0/F_0)$$

and since V_0 and F_0 are real and positive the zeros of Z are restricted to the negative real axis. Similarly the zeros of Y lie entirely on the negative real axis since $\hat{T}_0$ is zero in [9.35]. Thus both the poles and zeros of Z and Y are restricted to the negative real axis. Now from the Cauchy–Riemann equations we know that on the real axis

$$\frac{dZ}{ds} = \frac{dR}{d\sigma}\bigg|_{\omega=0} = \frac{dX}{d\omega}\bigg|_{\omega=0} \qquad [10.44]$$

where $Z = R+jX$, and $s = \sigma+j\omega$. Now

$$Z = R+jX = F_0 + \frac{V_0(\sigma-j\omega)}{\sigma^2+\omega^2}$$

and

$$X = \frac{-V_0\omega}{\sigma^2+\omega^2}$$

$$\frac{dX}{d\omega} = -\frac{\omega}{\sigma^2+\omega^2}\frac{\partial V_0}{\partial \omega} - \frac{(\sigma^2-\omega^2)V_0}{(\sigma^2+\omega^2)^2}$$

For $\omega = 0$

$$\frac{dX}{d\omega} = -\frac{V_0}{\sigma^2} = \frac{dR}{d\sigma} < 0 \qquad [10.45]$$

Thus the slope of an RC impedance along the negative real axis is negative. Using the same arguments as for the LC case we see that if the slope is not to change sign two zeros cannot follow each other along the axis, nor can two poles. Thus poles and zeros are *interlaced*. Similarly the zeros must be simple since otherwise $dZ/d\sigma$ would be zero there in contradiction to [10.45]. For the admittance (from [9.35])

$$Y = \hat{F}_0 + s\hat{V}_0 = G+jB \qquad [10.46]$$

and again on the real axis

$$\frac{dY}{ds} = \frac{dG}{d\sigma} = \frac{dB}{d\omega}$$

Now

$$B = \omega \hat{V}_0$$

and
$$\frac{dB}{d\omega} = \hat{V}_0 + \omega \frac{\partial \hat{V}_0}{\partial \omega}$$
$$= \hat{V}_0 > 0 \quad \text{when} \quad \omega = 0$$

Thus
$$\frac{dG}{d\sigma} = \hat{V}_0 > 0 \qquad [10.47]$$

on the real axis. Note that the slope of the impedance is negative while the slope of the admittance is positive along the real axis and as before since the slope of Y is positive its zeros must be simple as otherwise the slope would be zero there. Thus both the poles and zeros of Z (or Y) are simple, lie on the negative real axis and are interlaced.

Let us now consider the behaviour at zero and infinite frequencies. At zero frequency the capacitors become open circuits so that the network is composed of positive resistors and open circuits and hence the impedance $Z(0)$ is either infinite or equal to a finite real positive value R_0. Also $Y(0)$ is either zero or a finite real positive value G_0. When the frequency is infinite the capacitors become short circuits and the network is composed of positive resistors and short circuits so that $Z(\infty) = 0$ or a finite real positive value R_∞. Also $Y(\infty)$ is infinite or a finite real positive value G_∞. We may therefore write

$$Z(s) = \frac{\prod_1^m (s+\sigma_i)}{\prod_1^n (s+\sigma_j)} \qquad [10.48]$$

where the σ_i and σ_j are real, positive and distinct, and are interlaced. If $Z(0)$ is infinite one of the σ_j is zero to give a pole at the origin. If $Z(\infty) = R_\infty$ then $m = n$ and if $Z(\infty) = 0$ then $n = 1+m$. Typical plots of $Z(s)$ and $Y(s)$ are shown in Fig. 10.30.

Now $Z(s)$ may clearly be expanded in partial fractions to give

$$Z(s) = R_\infty + \frac{k_0}{s} + \sum_1^n \frac{k_j}{s+\sigma_j} \qquad [10.49]$$

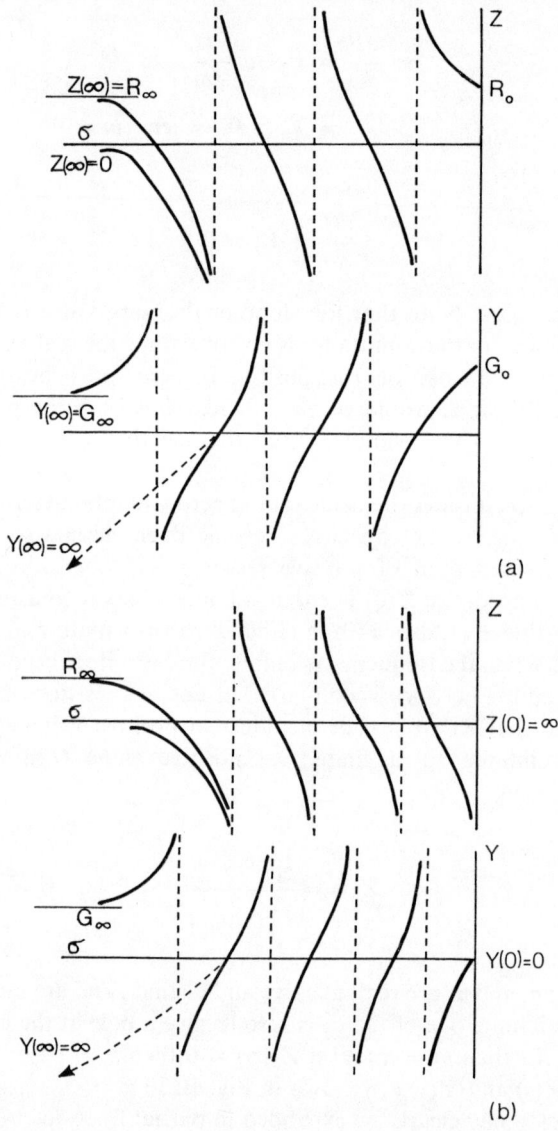

Fig. 10.30. *RC impedance functions* (a) *Impedance finite at the origin* (b) *Impedance infinite at the origin.*

with $R_\infty = 0$ if $Z(\infty) = 0$ and $k_0 = 0$ if $Z(0)$ is finite. The k_j are positive since the sign of the residue is the sign of the slope of $Y(s)$ at that point as given in [*10.12*] and the slope of Y is always positive. It follows from [*10.49*] that

$$Z(0) > Z(\infty) \qquad [10.50]$$

since

$$Z(\infty) = 0 \text{ or } R_\infty$$

and

$$Z(0) = R_\infty + \sum \frac{k_j}{\sigma_j} \text{ or infinity.}$$

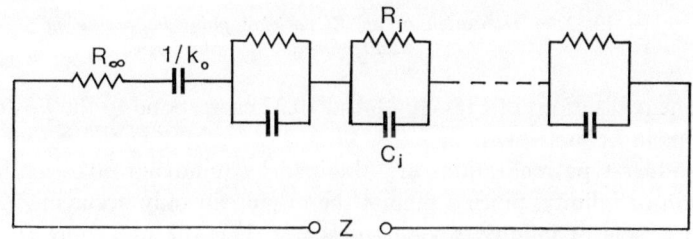

FIG. 10.31. *A realisation of an RC driving point impedance in series form.*

From [*10.49*] a realisation of $Z(s)$ is shown in Fig. 10.31 with

$$\sigma_j = \frac{1}{R_j C_j}$$

$$k_j = \frac{1}{C_j} \qquad [10.51]$$

In a similar fashion one may write a partial fraction expansion of $Y(s)$. Since $n \geq m$ we form an expansion of $Y(s)/s$ in the same form as $Z(s)$ so that

$$Y(s) = G_0 + \alpha_\infty s + \sum_1^m \frac{\alpha_i s}{s + \sigma_i} \qquad [10.52]$$

If $Y(0) = 0$ then $G_0 = 0$, if $Y(\infty) = G_\infty$ then $\alpha_\infty = 0$. All the α_i are positive since they are the residues of $Y(s)/s$ whose sign is determined by the slope of $sZ(s)$ = slope of $Z(s)$ times $-\sigma$. Fig. 10.32 is a realisation of $Y(s)$ according to [10.52] in which

$$\sigma_i = \frac{1}{R_i C_i}$$

$$\alpha_i = \frac{1}{R_i}$$

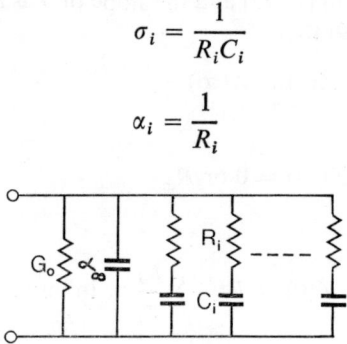

Fig. 10.32. *A realisation of an RC driving point impedance in parallel form.*

The realisations of Figs. 10.31 and 10.32 correspond to the Foster forms in *LC* networks.

Ladder type realisations are obtained by removing poles at the origin or infinity. Since a pole at the origin can only occur in $Z(s)$ and a pole at infinity can only occur in $Y(s)$ the procedure is to alternately remove resistors and capacitors.

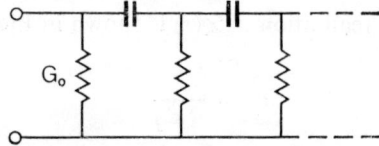

Fig. 10.33. *A realisation of an RC driving point impedance by extraction of poles at the origin.*

For example if we wish to remove poles at the origin we start with $Y(s)$ as in [10.52] and remove G_0 to leave a remainder with a zero at the origin. This corresponds to a pole in Z which is removed as a series capacitor. The cycle is repeated on the remainder and continued until the complete network has been synthesised as shown in Fig. 10.33. In the process for removing poles at infinity we start with $Z(s)$ as in [10.49] and remove R_∞ to leave a zero at infinity in the remainder. This corresponds to a pole in Y which is removed as a shunt capacitor. The cycle is repeated on the remainder and the process continued until the entire function is realised as shown in Fig. 10.34.

These procedures are best illustrated by an example. Let

$$Z(s) = \frac{s^3+6s^2+8s+2}{s^3+3s^2+2s} \qquad [10.54]$$

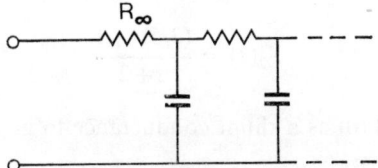

FIG. 10.34. *A realisation of an RC driving point impedance by extraction of poles at infinity.*

and suppose we remove poles at the origin. $Z(s)$ has such a pole so we may remove it directly. The residue at this pole is $sZ(s)|_{s=0} = 1$ so that the remainder is

$$Z_1(s) = Z(s) - (1/s)$$

$$= \frac{s^2+5s+5}{s^2+3s+2}$$

or

$$Y_1(s) = \frac{s^2+3s+2}{s^2+5s+5}$$

We then remove a shunt conductance $Y(0)$ to create a zero at the origin in the remainder. Thus

$$Y_2(s) = Y_1(s) - (2/5)$$

$$= \frac{(3/5)s^2+s}{s^2+5s+5}$$

and

$$Z_2(s) = \frac{s^2+5s+5}{(3/5)s^2+s}$$

We remove the pole at the origin as a series capacitor so that

$$Z_3(s) = Z_2(s) - (5/s)$$

$$= \frac{s+2}{(3/5)s+1}$$

Then

$$Y_3(s) = \frac{(3/5)s+1}{s+2}$$

and we remove $Y(0)$ as a shunt conductance to give

$$Y_4(s) = Y_3(s) - (1/2)$$

$$= \frac{s/10}{s+2}$$

and the complete network is that shown in Fig. 10.35.

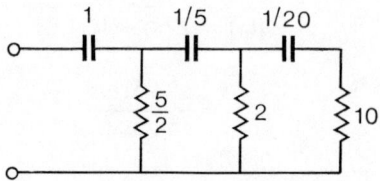

FIG. 10.35. *Realisation of the impedance given in [10.54] by extraction of poles at the origin.*

Let us now realise the same impedance by means of a network of the type shown in Fig. 10.34. We first remove a series resistor equal to $Z(\infty)$ so that

$$Z_1(s) = Z(s) - 1$$

$$= \frac{3s^2 + 6s + 2}{s^3 + 3s^2 + 2s}$$

and

$$Y_1(s) = \frac{s^3 + 3s^2 + 2s}{3s^2 + 6s + 2}$$

TWO-ELEMENT-KIND SYNTHESIS

The residue at infinity is 1/3 and we remove a shunt capacitor to give

$$Y_2(s) = Y_1(s) - (s/3)$$
$$= \frac{2s^2 + (4/3)s}{3s^2 + 6s + 2}$$

and

$$Z_2(s) = \frac{3s^2 + 6s + 2}{2s^2 + (4/3)s}$$

We then remove a series resistor equal to $Z(\infty)$ to give

$$Z_3(s) = Z_2(s) - 3/2$$
$$= \frac{4s + 2}{2s^2 + (4/3)s}$$

and

$$Y_3(s) = \frac{2s^2 + (4/3)s}{4s + 2}$$

Removing the pole at infinity as a shunt capacitor gives

$$Y_4(s) = Y_3(s) - (1/2)s$$
$$= \frac{(1/3)s}{4s + 2}$$

and the complete network is shown in Fig. 10.36.

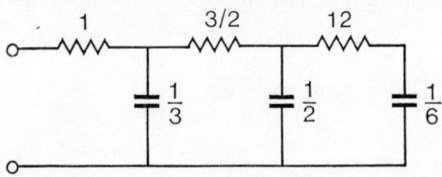

FIG. 10.36. *Realisation of the impedance given in [10.54] by extraction of poles at infinity.*

We thus see that an *RC* realisation exists corresponding to each of the four basic forms of *LC* network. Again combinations of these basic forms can be used to generate a wide variety of possible realisations for any given impedance.

All the properties of RC networks can be derived from the corresponding properties of LC networks by means of a simple transformation. To see this we recall that any LC driving point impedance may be written in accordance with [10.7] as

$$Z_{LC}(s) = k_\infty s + \frac{k_0}{s} + \sum \frac{2k_i s}{s^2 + \omega_i^2} \qquad [10.55]$$

while any RC driving point impedance may be written in accordance with [10.49] as

$$Z_{RC}(s) = R_\infty + \frac{\beta_0}{s} + \sum \frac{\beta_j}{s + \sigma_j} \qquad [10.56]$$

with a change of notation in which k is replaced by β.

Now to obtain [10.56] from [10.55] we may carry out the following steps:

(i) divide Z_{LC} by s which gives

$$\frac{Z_{LC}(s)}{s} = k_\infty + \frac{k_0}{s^2} + \sum \frac{2k_i}{s^2 + \omega_i^2} \qquad [10.57]$$

(ii) replace s^2 by s, and ω_i^2 by σ_j and call the result $Z_{RC}(s)$.

All the properties of Z_{RC} are then obvious from the observations that k_∞, k_0 and k_i are positive and the ω_i^2 are positive. In circuit terms the above process replaces each inductor L by a resistor $R = L$ and leaves each capacitor unchanged. Thus if we know all the properties and processes of LC networks the corresponding ones for RC networks follow immediately. In symbolic terms the transformation can be written as

$$Z_{RC}(p) = \left\{\frac{1}{s} Z_{LC}(s)\right\}\bigg|_{s^2 = p} \qquad [10.58]$$

where p is then used as the complex frequency variable for the RC network. Similarly

$$\frac{1}{s} Z_{LC}(s) = Z_{RC}(p)\bigg|_{p=s^2} \qquad [10.59]$$

A similar argument relating to LC and RC admittances shows that

$$Y_{RC}(p) = \{s Z_{LC}(s)\}\big|_{s^2 = p} \qquad [10.60]$$

and
$$sY_{LC}(s) = Y_{RC}(p)\big|_{p^2=s}$$

Thus for example if we wish to realise the RC impedance function given in [10.54], i.e.

$$Z_{RC}(p) = \frac{p^3+6p^2+8p+2}{p^3+3p^2+2p}$$

we could first form the corresponding LC impedance using [10.58] as

$$\frac{1}{s}Z_{LC}(s) = \frac{s^6+6s^4+8s^2+2}{s^6+3s^4+2s^2}$$

so that

$$Z_{LC}(s) = \frac{s^7+6s^5+8s^3+2s}{s^6+3s^4+2s^2}$$

and then synthesise $Z_{LC}(s)$ by any of the methods given earlier for LC networks. Each inductor L in the resulting circuit is replaced by a resistor $R = L$ and each capacitor left unchanged to find the realisation of $Z_{RC}(p)$.

10.4. RL Driving Point Impedances

RL networks are of much less importance than the LC or RC class. They have little practical importance and all their properties can be derived from LC networks by a simple transformation of the type given in [10.58]. In fact

$$Z_{RL}(p) = \{sZ_{LC}(s)\}\big|_{s^2=p} \qquad [10.61]$$

In circuit terms each capacitor C is replaced by a resistor $R = 1/C$ and each inductor is left unchanged. The reverse transformation is

$$sZ_{LC}(s) = Z_{RL}(p)\big|_{p=s^2} \qquad [10.62]$$

Thus the properties of an RL impedance are the same as those of an RC admittance and the properties of an RL admittance are the same as those of an RC impedance. RL networks will therefore not be considered separately in the remainder of this text.

10.5. Transfer Function Synthesis without Ideal Transformers

In Section 10.2 we have seen that the synthesis of LC transfer functions in general, requires the use of ideal transformers. Because of the relationships given in the transformations of [*10.58*] and [*10.61*] this is also true for *RL* and *RC* networks. From a practical viewpoint this is a grave disadvantage particularly in the case of *RC* networks whose chief merit is that they avoid the use of coils of any description. There is therefore considerable interest in the synthesis of two-element-kind two-ports without transformers. It should be said immediately that it is not known up to now how to realise an arbitrary two-port *z*-matrix without transformers in the way that has been shown in

FIG. 10.37. *An open-circuit RC two-port.*

Section 10.2 for the case with transformers. However, in *RC* networks there is often a need for the realisation of a transfer function alone. Such networks are often driven by sources of negligible internal impedance and are terminated in loads which may effectively be regarded as open-circuits. This is illustrated in Fig. 10.37. The quantity of interest is then the open-circuit voltage transfer ratio

$$T_{12} = E_2/E_1$$
$$= 1/A \qquad [10.63]$$
$$= -y_{12}/y_{22}$$

Now if T_{12} is specified as

$$T_{12} = N/D \qquad [10.64]$$

we divide numerator and denominator by a polynomial Q so that

$$T_{12} = (N/Q)/(D/Q) \qquad [10.65]$$

and we equate

$$-y_{12} = N/Q$$
$$y_{22} = D/Q \qquad [10.66]$$

It is clear, therefore, that since y_{22} is the driving point admittance of an *RC* network, Q must be chosen such that it is the product of

TWO-ELEMENT-KIND SYNTHESIS   435

factors with real negative zeros which interlace appropriately with those of D. It also follows that a necessary condition on T_{12} is that its poles be real, negative and distinct. The problem therefore can be reduced to the simultaneous synthesis of y_{22} and $-y_{12}$ as given in [10.66]. Note that y_{11} is left unspecified and this enables some success to be achieved in synthesis without transformers. There are very many techniques known for tackling this problem and a complete treatment is beyond the scope of this book, so that only an outline of one or two methods will be given.

Complete necessary and sufficient conditions on T_{12} are known and a realisation procedure is available. (See D. Hazony, 1963. *Elements of Network Synthesis*, Reinhold Publishing Corp., New York.) The resulting networks, however, are not very desirable from a practical viewpoint and considerable attention has been devoted to seeking methods whereby T_{12} may be realised within a constant multiplier which is unknown before the realisation is completed. Thus if T_{12} is specified we shall be content to realise a function KT_{12} where K is unspecified. This is equivalent to realising y_{22} exactly and $-y_{12}$ within an arbitrary multiplier.

10.5.1. TRANSMISSION ZEROS AT THE ORIGIN AND INFINITY

Now y_{22} is an *RC* driving-point admittance and $-y_{12}$ has the same poles as y_{22}. We therefore tend to classify the problem according to the zeros of $-y_{12}$ which are also the zeros of T_{12} and are known as the *Zeros of Transmission*. If the numerator of T_{12} is either constant or is s^r, synthesis is particularly simple. These provide zeros of transmission at the origin and infinity. Considering first the case where the zeros of transmission are all at infinity the procedure is merely to realise y_{22} as a ladder network with series resistors and shunt capacitors. The resulting y_{12} will have the same poles as y_{22} and zeros of transmission only at infinity whose number is equal to the number of capacitors. This occurs because if we take a general ladder such as that shown in Fig. 10.38 then we obtain a zero of transmission when any series branch is an open-circuit and when any shunt branch is a short-circuit since in either case no signal can travel through the network. Thus in the present case the shunt capacitors become short-circuits at infinity and each contributes one zero of transmission. This method clearly realises y_{22} exactly and

F

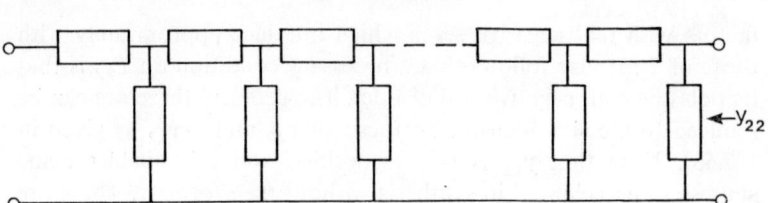

Fig. 10.38. *A general ladder network.*

$-y_{12}$ with the correct poles and zeros but an unknown multiplying constant whose value is determined by evaluating $y_{12}(0)$. Notice that impedance scaling does not affect the value of the multiplier on T_{12} since it acts on both y_{22} and $-y_{12}$ and leaves their ratio unchanged.

The case where all the zeros of transmission are at the origin is obviously realised by synthesising y_{22} in the form of a ladder with series capacitors and shunt resistors.

In choosing the polynomial Q in [10.65] we should ensure that it is such as to provide a realisation where the first element at port 2 is a shunt component and the first element at port 1 is a series component since an initial series element at port 2 or shunt element at port 1 does not affect T_{12}.

To illustrate these procedures let us take

$$T_{12} = \frac{K}{(s+1)(s+3)}$$

so that

$$-y_{12} = K/Q$$

$$y_{22} = \frac{(s+1)(s+3)}{Q}$$

Now since the transmission zeros are all at infinity we shall realise y_{22} with shunt capacitors and series resistors, and since we require the first element at port 2 to be a shunt element it should be a capacitor so we must arrange that y_{22} has a pole at infinity. Thus Q is chosen to have degree 1. So we put

$$Q = s+2$$

We then realise y_{22} as

$$y_{22} = s + \frac{2s+3}{s+2}$$

$$= s + y_1$$

$$z_1 = \tfrac{1}{2} + \frac{\tfrac{1}{2}}{s+2}$$

$$= \tfrac{1}{2} + z_2$$

$$y_2 = 2s + 4$$

leading to the circuit of Fig. 10.39. Note that although T_{12} has two zeros of transmission at infinity as required, y_{12} has only one.

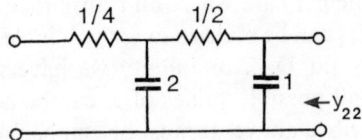

FIG. 10.39. *Realisation of an open-circuit voltage transfer function by an RC network.*

A similar procedure is adopted when the zeros of transmission lie at the origin. To find the value of K we evaluate T_{12} at the origin or at infinity. In the above example $T_{12}(0) = 1$ so that $K = 3$. By a slight modification of the circuit in Fig. 10.39, it is possible to control the value of K such that $T_{12}(0)$ can take any value between 0 and 1. The modified circuit, in general, is that shown in Fig. 10.40. $G_1 + G_2$ is maintained at the correct value to realise the chosen y_{22} (remember that to evaluate y_{22} we short-circuit E_1) and the value of $T_{12}(0)$ is

$$T_{12}(0) = \frac{G_1}{G_1 + G_2}$$

which can be varied from 0 to 1 by varying G_1/G_2. A similar procedure with capacitors rather than resistors can be used in the case where the zeros of transmission are all at the origin to vary $T_{12}(\infty)$ between 0 and 1.

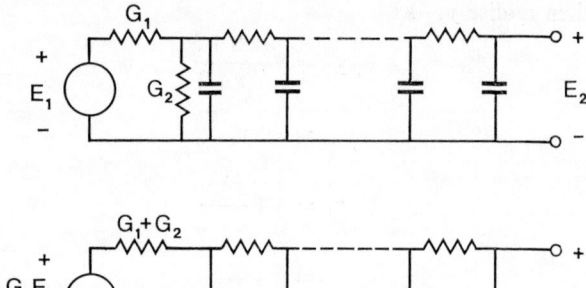

Fig. 10.40. *Circuit for controlling $T_{12}(0)$ when all transmission zeros are at infinity.*

Thus in the particular case where all transmission zeros are at the origin or infinity T_{12} can be realised *exactly*. It is clear that $T_{12}(0)$ can never exceed unity (at D.C. or infinity we have a purely resistive network), so that all possible values of K can be accommodated by the technique described above. It must be emphasised that this is not possible in general.

In the case of *LC* networks if all the zeros of transmission lie at the origin or at infinity any two parameters can be realised exactly by the same technique and we shall use this technique in later realisations of certain *RLC* transfer functions.

10.5.2. Transmission zeros on the negative real axis

Returning to *RC* networks and the general ladder structure of Fig. 10.38 we note that this is capable of realising zeros of transmission only when such zeros lie on the negative real axis (including the origin and infinity) since a transmission zero occurs when the impedance or admittance of a branch is infinite. But each branch is an *RC* impedance and we know that the poles of such impedances or admittances are restricted to the negative real axis. If, therefore, T_{12} has zeros at points on the negative real axis we must attempt to realise the resulting y_{22} such that it has a series branch which is open-circuit or a shunt branch which is short-circuit at each of the prescribed transmission zeros. Thus if we begin to realise y_{22} in the situation where none of the transmission zeros is a pole of y_{22} we must remove some shunt element so as to create a remainder which has a zero at one of the transmission zeros, and which is still an RC admittance. In that case the remaining impedance has a pole at the transmission zero which can be removed as a series circuit whose impedance is infinite at the transmission zero. This process may then be repeated until all the transmission zeros are realised. The process is best understood by means of an example. Let

$$T_{12} = \frac{K(1+s)(\tfrac{1}{2}+s)}{2s^2+9s+8}$$

and choose $Q = 4s+5$ so that

$$Y_{22} = \frac{2s^2+9s+8}{4s+5}$$

We must now remove a shunt element from y_{22} so as to create a zero at one of the transmission zeros, say at $s = -1$. Now we have a choice of whether to remove a resistor, a capacitor or partially remove a finite pole but in any case we must ensure that the remainder is still an RC admittance. Thus if we remove a conductance its value must not exceed the constant term in the partial fraction expansion of y_{22}, if we remove a capacitance its value must not exceed the residue at infinity and if we partially remove a pole the residue must not exceed the residue of y_{22} at that pole. In the present case let us remove a conductance such that $y_{22}-g_1 = 0$ at $s = -1$. This gives $g_1 = 1$ and a remainder

$$y_1 = \frac{2s^2+5s+3}{4s+5}$$

or

$$z_1 = \frac{4s+5}{(s+1)(2s+3)}$$

We then remove the pole at $s = -1$ from z_1 as a series branch to give

$$z_2 = z_1 - \frac{1}{s+1}$$

$$= \frac{2}{2s+3}$$

or

$$y_2 = \frac{2s+3}{2}$$

We must now remove a parallel branch from y_2 to create a zero in the remainder at the other transmission zero, $s = -\frac{1}{2}$. Again we try a conductance such that $y_2 - g_2 = 0$ at $s = -\frac{1}{2}$ which gives $g_2 = 1$ and

$$y_3 = y_2 - 1$$
$$= \frac{2s+1}{2}$$

or

$$z_2 = \frac{2}{2s+1}$$

which is realised as a series branch to give the circuit of Fig. 10.41. The value of K can be found by calculating $T_{12}(\infty)$ from Fig. 10.41. This gives $T_{12}(\infty) = 1$ so $K = 2$.

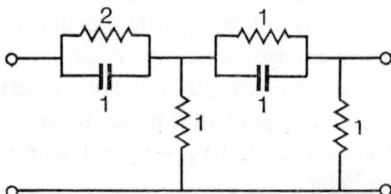

Fig. 10.41. *Realisation of an open-circuit voltage transfer ratio with zeros of transmission on the negative real axis.*

The dual of this procedure can, of course, also be carried out and we would then realise the transmission zeros by shunt branches and use the series branches to produce poles of admittance at the required values of s. Indeed the two methods may be mixed and, since, in addition, we have a choice of the type of element removed in creating the pole (R, C, or partial pole), it is obvious that there are a great many possibilities. However, there is no guarantee that any of the possible combinations will be successful in a particular case, since it has not been shown that y_{22} may be realised by this form of network with prescribed series or shunt branches. In practice therefore, one must try all sorts of combinations (including varying the order in which one attempts to realise the transmission zeros) to try to find a realisation. It is, however, worthwhile to do so since if the method succeeds it is usually the most economical of those available.

10.5.3. GUILLEMIN'S METHOD

In the following realisation procedure, due to E. A. Guillemin, it is possible to realise any prescribed T_{12} whose numerator is a polynomial with *positive* coefficients, in the form of a parallel connection of ladder networks. Let

$$T_{12} = \frac{\Sigma a_i s^i}{D(s)} \qquad [10.67]$$

where all the a_i are positive. Again we choose a polynomial $Q(s)$ such that $D(s)/Q(s)$ is an RC admittance and we write

$$T_{12} = \frac{\{\Sigma a_i s^i\}/Q(s)}{D(s)/Q(s)} \qquad [10.68]$$

and put

$$y_{22} = D(s)/Q(s) \qquad [10.69]$$

$$-y_{12} = \{\Sigma a_i s^i\}/Q(s) \qquad [10.70]$$

This pair is then realised as a parallel connection of networks each having y_{22} as given in [10.69] and the ith network has

$$-y_{12}^{(i)} = a_i s^i / Q(s)$$

Thus the ith network has i zeros of transmission at the origin and $n-i$ at infinity, where n is the degree of $Q(s)$. The realisation therefore consists of realising y_{22} with i series capacitors and $n-i$ shunt capacitors thus yielding $K_i y_{12}^{(i)}$. The network is then scaled to produce $y_{12}^{(i)}$ and y_{22}/K_i. These networks are then connected in parallel to give y_{12} and $y_{22}[\Sigma(1/K_i)]$ and thus realise T_{12} within an arbitrary multiplier. Clearly it is only necessary for the a_i to be positive to achieve this realisation. As an example consider

$$T_{12} = \frac{s^2+s+1}{s^2+4s+3}$$

and choose $Q = (s+2)(s+4)$. Thus

$$y_{22} = \frac{s^2+4s+3}{(s+2)(s+4)}$$

and the $y_{12}^{(i)}$ are

$$-y_{12}^{(1)'} = \frac{s^2}{s^2+6s+8}$$

$$-y_{12}^{(2)} = \frac{s}{s^2+6s+8}$$

$$-y_{12}^{(3)} = \frac{1}{s^2+6s+8}$$

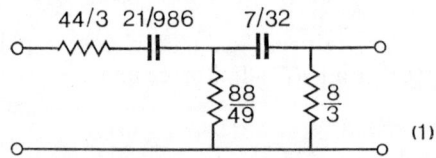

(1)

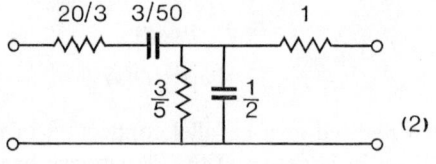

(2)

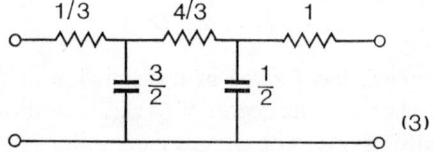

(3)

FIG. 10.42. *Realisation of an open-circuit voltage transfer ratio by Guillemin's procedure.*

The three networks are realised as shown in Fig. 10.42 and the multipliers are calculated from the realisations to give $K_1 = 3/44$, $K_2 = 3/10$, $K_3 = 8$. The networks are then impedance scaled to produce the exact values of the y_{12}^i and then connected in parallel. Thus in network (1) the impedance level is raised by a factor 3/44, in

(2) by a factor 3/10 and in (3) by a factor 8. Thus the final realisation gives the exact value of y_{12} and y_{22} is multiplied by $(\frac{44}{3}+\frac{10}{3}+\frac{1}{8}) = 145/8$ and so T_{12} has a multiplier 8/145.

This method of synthesis is simple and easy to implement but employs a relatively large number of elements and usually results in a multiplier on T_{12} which is rather small. There are many other possible methods of realisation and indeed variations on the methods presented here are quite common. The whole area is one of considerable specialisation and the reader should consult one of the several excellent texts on this subject, e.g. the work by Hazony mentioned at the beginning of Section 10.5.

10.6. Concluding Remarks

In this chapter we have been concerned with the synthesis of two-element kind driving point and transfer functions. The main features are that any such driving-point impedance may be realised in a number of ways without ideal transformers while the realisation of a two-port specification requires, in general, ideal transformers and/or coupled coils. The fact that driving point impedance realisations do not require ideal transformers makes their implementation particularly simple and we shall see in the next chapter that for *RLC* networks ideal transformers are necessary for both driving-point and transfer function realisations.

Another notable feature of two-element-kind networks is that when the zeros of transmission are all at infinity or the origin exact transfer function synthesis without ideal transformers is possible. This will be found to have important consequences in the synthesis of *RLC* networks.

Problems

10.1. For a reactance function show that $(dX/d\omega) > (X/\omega)$ making direct use of [*10.3*].

10.2. Synthesise each of the following functions as the driving-point *impedance* of an *LC* network using each of the Foster forms in turn

(a) $\dfrac{s(s^2+2)(s^2+4)}{(s^2+1)(s^2+3)(s^2+5)}$

(b) $\dfrac{(s^2+2)(s^2+4)}{s(s^2+3)(s^2+5)}$

(c) $\dfrac{s(s^2+2)(s^2+4)}{(s^2+1)(s^2+3)}$

10.3. Realise each of the functions given in Problem 10.2 as the driving-point *admittance* of an *LC* network using each of the Foster forms in turn.

10.4. Two impedance functions Z_1 and Z_2 have zeros at $s = 0$, $s = \pm j1$, $s = \pm j2$ and the slopes of the functions at these zeros are 1, 2, 3, and 2, 3, 4, respectively. Show that the admittance $Y_1 - Y_2$ is a reactance function and find a network to realise it.

10.5. Synthesise the impedance function given below in the form shown in Fig. 10.43

$$Z = \frac{(s^2+1)(2s^4+13s^2+4)}{5s^5+31s^3+12s}$$

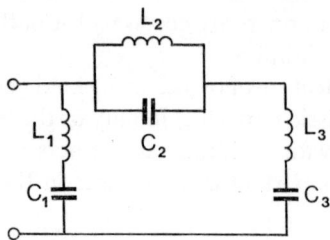

FIG. 10.43. *Problem 10.5.*

10.6. Synthesise each of the functions given in Problem 10.2 as the driving point *impedance* of an *LC* network using the Cauer form with series inductors and shunt capacitors.

10.7. Realise each of the functions given below as the driving-point impedance of an *LC* network using the Cauer form with series capacitors and shunt inductors.

(a) $\dfrac{s^5+3s^3+s}{3s^4+4s^2+1}$

(b) $\dfrac{10s^4+7s^2+1}{4s^5+5s^3+s}$

10.8. Realise each of the functions given in Problem 10.2 as the driving-point *admittance* of an LC network using the Cauer form with series inductors and shunt capacitors.

10.9. Realise the following two-port impedance parameters by a lossless network, using coupled coils in place of ideal transformers where possible.

(a) $z_{11} = \dfrac{s^2+2}{s}$; $z_{12} = \dfrac{2s^2+2}{s}$; $z_{22} = \dfrac{4s^2+2}{s}$

(b) $z_{11} = \dfrac{2s^2+2}{s}$; $z_{12} = \dfrac{2s^2+2}{s}$; $z_{22} = \dfrac{4s^2+2}{s}$

(c) $z_{11} = \dfrac{s^2+2}{s}$; $z_{12} = \dfrac{2s^2-2}{s}$; $z_{22} = \dfrac{4s^2+2}{s}$

(d) $z_{11} = \dfrac{s^2+2}{s}$; $z_{12} = \dfrac{2-2s^2}{s}$; $z_{22} = \dfrac{4s^2+2}{s}$

(This is a Darlington C-section)

(e) $z_{11} = \dfrac{s^4+3s^2+1}{s^3+s}$; $z_{12} = \dfrac{2s^4+2s^2+1}{s^3+s}$; $z_{22} = \dfrac{4s^4+6s^2+1}{s^3+s}$

(This is a Darlington D-section).

10.10. Realise the following two-port admittance parameters by a lossless network, using coupled coils in place of ideal transformers where possible.

(a) $y_{11} = \dfrac{s^2+1}{s}$; $-y_{12} = \dfrac{s^2+2}{s}$; $y_{22} = \dfrac{s^2+4}{s}$

(b) $y_{11} = \dfrac{s^2+4}{s}$; $\quad -y_{12} = \dfrac{s^2-2}{s}$; $\quad y_{22} = \dfrac{s^2+1}{s}$

(This is another version of the Darlington C-section)

(c) $y_{11} = \dfrac{s^4+4s^2+1}{s^3+s}$; $\quad -y_{12} = \dfrac{s^4+s^2+2}{s^3+s}$; $\quad y_{22} = \dfrac{s^4+7s^2+4}{s^3+s}$

(This is another version of the Darlington D-section).

10.11. Synthesise the following *RC* driving point *impedance* functions in (i) series form (ii) parallel form.

(a) $Z(s) = \dfrac{(s+4)(s+6)}{(s+1)(s+5)}$

(b) $Z(s) = \dfrac{(s+2)(s+4)}{s(s+3)}$

(c) $Z(s) = \dfrac{(s+4)(s+6)}{(s+1)(s+5)(s+7)}$

(d) $Z(s) = \dfrac{(s+2)(s+4)}{s(s+3)(s+5)}$

10.12. Synthesise the *RC* impedance functions (*a*) and (*b*) in ladder form with series capacitors and the *RC* impedance functions (*c*) and (*d*) in ladder form with shunt capacitors.

(a) $Z(s) = \dfrac{s^2+3s+1}{3s^2+4s+1}$

(b) $Z(s) = \dfrac{6s+2}{4s^2+5s+1}$

(c) $Z(s) = \dfrac{s^2+3s+1}{s^3+4s^2+3s}$

(d) $Z(s) = \dfrac{4s^2+10s+4}{2s^2+4s+1}$

TWO-ELEMENT-KIND SYNTHESIS

10.13. Convert each of the driving point impedance functions of Problem 10.2 into the corresponding RL impedance and give the corresponding network modification in each case. Repeat for the admittance functions of Problem 10.3.

10.14. Convert each of the RC driving point impedance functions in Problem 10.11 into corresponding RL impedance functions and give the required network modifications.

10.15. Synthesise the following pairs of RC admittance functions without ideal transformers in the form of single ladder structures.

(a) $y_{22} = \dfrac{(s+2)(s+4)}{s+3}$; $\quad -y_{12} = \dfrac{8}{s+3}$

(b) $y_{22} = \dfrac{(s+2)(s+4)}{s+3}$; $\quad -y_{12} = \dfrac{4}{s+3}$

(c) $y_{22} = \dfrac{(s+2)(s+4)}{s+3}$; $\quad -y_{12} = \dfrac{s^2}{s+3}$

(d) $y_{22} = \dfrac{(s+2)(s+4)}{s+3}$; $\quad -y_{12} = \dfrac{\frac{1}{2}s^2}{s+3}$

(e) $y_{22} = \dfrac{8s^2+20s+11}{12s+10}$; $\quad -y_{12} = \dfrac{2(s+1)(2s+1)}{6s+5}$

10.16. Synthesise the following pair of RC admittance functions by Guillemin's method, and determine the value of K.

$$y_{22} = K\dfrac{s^2+4s+3}{s^2+6s+8}; \quad -y_{12} = \dfrac{s^2+2s+3}{s^2+6s+8}$$

Chapter 11

SYNTHESIS OF RLC DRIVING-POINT IMPEDANCES

In the previous chapter the synthesis of two-element-kind driving-point impedances was discussed in some considerable detail, and the object of this chapter is to consider the corresponding problem for *RLC* networks. We know already from Chapter 9 that the necessary condition is that any impedance or admittance function must be positive real and we must now show that the condition is in fact sufficient.

Initially one naturally seeks to find out if the methods of the previous chapter are again sufficient. One can expand the impedance or admittance in partial fractions and then attempt to realise each term separately. Let

$$Z(s) = \sum \left(\frac{k_i}{s+s_i} + \frac{k_i^*}{s+s_i^*} \right) + \sum \frac{\alpha_i}{s+\sigma_i} \qquad [11.1]$$

where s_i are the complex and imaginary poles and σ_i are any real poles. Now we know from Chapter 9 that any terms corresponding to imaginary poles can always be realised since the positive real condition ensures that the residues at such poles are real and positive. In general we therefore need to consider the realisation of a term like

$$z_i = \frac{s(k_i+k^*)+k_i s_i^* + k_i^* s_i}{s^2+(s_i+s_i^*)s+s_i s_i^*} \qquad [11.2]$$

which requires a circuit of the type shown in Fig. 11.1. This may be synthesised from [11.2] by writing

$$y_i = \frac{1}{z_i}$$

$$= \frac{s^2+2\sigma_i s+(\sigma_i^2+\omega_i^2)}{2(\alpha_i s+\sigma_i \alpha_i+\beta_i \omega_i)}$$

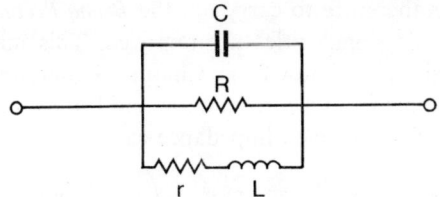

FIG. 11.1. *Realisation of the impedance given by [11.2].*

where
$$k_i = \alpha_i + j\beta_i$$
$$s_i = \sigma_i + j\omega_i$$

Now y_i may be written as

$$y_i = \frac{1}{2}\left[\frac{s}{\alpha_i} + \frac{1}{\alpha_i}\left(\sigma_i - \frac{\beta_i}{\alpha_i}\omega_i\right) + \frac{\omega_i^2(1+[\beta_i^2/\alpha_i^2])}{\alpha_i s + (\alpha_i\sigma_i + \beta_i\omega_i)}\right]$$

and the admittance of the circuit in Fig. 11.1 is

$$y = Cs + \frac{1}{R} + \frac{1}{r+Ls}$$

Thus for positive element values we require

$$\alpha_i \geqq 0$$
$$\alpha_i\sigma_i - \beta_i\omega_i \geqq 0 \qquad [11.3]$$
$$\alpha_i\sigma_i + \beta_i\omega_i \geqq 0$$

Now the positive real condition does not guarantee that these conditions will be satisfied so that, in general, an arbitrary positive real impedance function can *not* be realised by a partial fraction expansion. We must therefore investigate other methods of synthesis. Historically the realisation of an arbitrary positive real function was accomplished in 1930 by O. Brune and in the next section we shall describe his procedure.

11.1. The Brune Procedure

If we are given an arbitrary positive real impedance function then in general it may have poles or zeros on the imaginary axis. The Brune procedure as such applies to functions which are devoid of poles or zeros on the imaginary axis, including the origin and infinity.

450 CIRCUIT THEORY

The first step is therefore to carry out the *Brune Preamble* in which all imaginary axis singularities are removed. This may be readily accomplished since we know from Chapter 9 that we may always extract a *j*-axis pole from a p.r. function and leave a p.r. remainder. We therefore write the given impedance as

$$Z(s) = Z_1(s) + \sum \frac{2k_i s}{s^2 + \omega_i^2} + \left(k_\infty s + \frac{k_0}{s}\right) \qquad [11.4]$$

where $Z_1(s)$ is devoid of *j*-axis poles. The final bracketed terms appear only if there is a pole at the origin or infinity. The situation corresponding to [11.4] is shown in Fig. 11.2(*a*). We then write

$$Y_1(s) = Y_2(s) + \sum \frac{2\alpha_j s}{s^2 + \omega_j^2} + \left(\alpha_\infty s + \frac{\alpha_0}{s}\right)$$

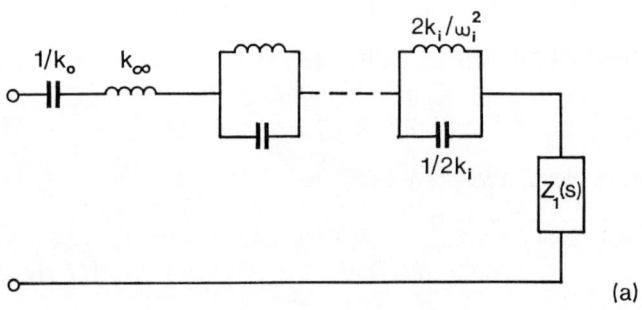

(a)

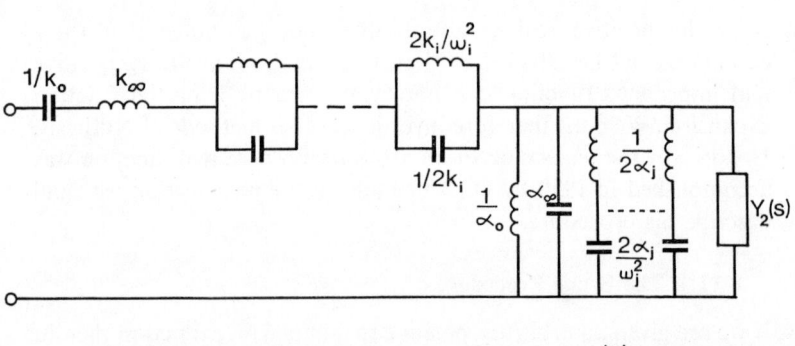

(b)

Fig. 11.2. (a) *Realisation of imaginary-axis impedance poles in the Brune Preamble;* (b) *Realisation of imaginary axis admittance poles in the Brune Preamble.*

SYNTHESIS OF RLC DRIVING-POINT IMPEDANCES

where $Y_2(s)$ is devoid of j-axis poles. The network representation is shown in Fig. 11.2(b). By these means we may clearly end up with an impedance function, devoid of j-axis poles and zeros, having equal degrees in numerator and denominator, and the constant term being present in both. The latter two conditions are true because there are no poles or zeros at the origin or infinity. A function having the above properties is called a minimum reactance function (no poles of z on the imaginary axis) and a minimum susceptance function (no poles of y on the imaginary axis).

Having completed the Brune Preamble the next step is to examine the real part of the remaining impedance $Z(j\omega)$. Now if

$$Z(s) = \frac{m_1 + n_1}{m_2 + n_2} \qquad [11.5]$$

With $m_{1,2}$ even and $n_{1,2}$ odd then the even part of $Z(s)$ is

$$evZ(s) = \frac{m_1 m_2 - n_1 n_2}{m_2^2 - n_2^2} \qquad [11.6]$$

and

$$\operatorname{Re} Z(j\omega) = evZ(s)|_{s=j\omega} \qquad [11.7]$$

A plot of $\operatorname{Re} Z(j\omega)$ might be of the form given in Fig. 11.3 and at some frequency ω_1 it achieves its absolute minimum value, R_1. We

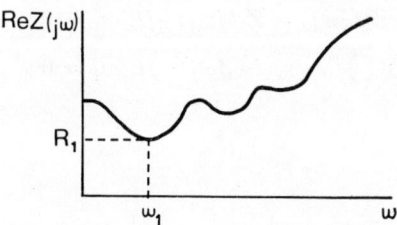

FIG. 11.3. *Real part of a driving-point impedance at real frequencies.*

may therefore extract any series resistance up to the value R_1, from $Z(s)$, and still leave a positive real remainder since by doing so the poles are unaffected and the real part of the remainder for $s = j\omega$ will be non-negative. In fact if we write

$$Z_1(s) = Z(s) - R_1 \qquad [11.8]$$

then $Z_1(s)$ is positive real and

$$Z_1(j\omega_1) = jX_1(\omega_1) \qquad [11.9]$$

i.e. $Z_1(j\omega_1)$ is purely imaginary.

The next step depends upon whether $X_1(\omega_1)$ is positive or negative. The object is to reduce $Z_1(j\omega_1)$ to zero. This is done by extracting a series inductance L_1 from $Z_1(s)$ to give

$$Z_2(s) = Z_1(s) - L_1 s \qquad [11.10]$$

But since $Z_1(s)$ does not have a pole at infinity (numerator and denominator degrees are equal) $Z_2(s)$ does have a pole at infinity and since $Z_2(s) \to -L_1 s$ as s tends to infinity the residue is $-L_1$. Thus if L_1 is positive this residue is negative and $Z_2(s)$ is not positive real. If L_1 is negative then the residue is positive and $Z_2(s)$ is positive real (and, of course, of higher degree than $Z_1(s)$!). We are thus faced with two rather unattractive alternatives. Now, we certainly cannot proceed if $Z_2(s)$ is not positive real since we are attempting to find a realisation for an impedance function which *is* positive real. On the other hand we might proceed by extracting a negative inductance, leaving a remainder of increased degree, and hope to absorb the negative inductance and reduce the degree later. If this is successful we may attain our ends since at least the remainder is positive real.

The value of L_1 is chosen so that

$$Z_2(j\omega_1) = Z_1(j\omega_1) - jL_1\omega_1 = 0$$
$$= jX_1(\omega_1) - jL_1\omega_1 = 0$$

or

$$L_1 = \frac{X_1(\omega_1)}{\omega_1} \qquad [11.11]$$

But since we require $L_1 < 0$ this extraction can only be used when $X_1(\omega_1) < 0$. This step is illustrated in Fig. 11.4. Now, since $Z_2(s)$ is zero at $s = \pm j\omega_1$ and is positive real, $Y_2(s)$ has a pole at $s = \pm j\omega_1$ and is also positive real. Hence from Chapter 9, Section 9.21 (vii) we may extract this pole and leave a positive real remainder. Thus

$$Y_3(s) = Y_2(s) - \frac{2ks}{s^2 + \omega_1^2} \qquad [11.12]$$

SYNTHESIS OF RLC DRIVING-POINT IMPEDANCES

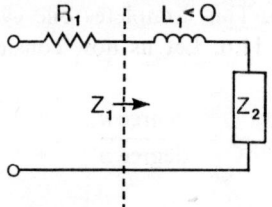

FIG. 11.4. *First stage in the Brune Synthesis Procedure when $X_1(\omega_1) < 0$.*

and this is shown in Fig. 11.5 from which

$$2k = \frac{1}{L_2}$$

$$\omega_1^2 = \frac{1}{L_2 C} \qquad [11.13]$$

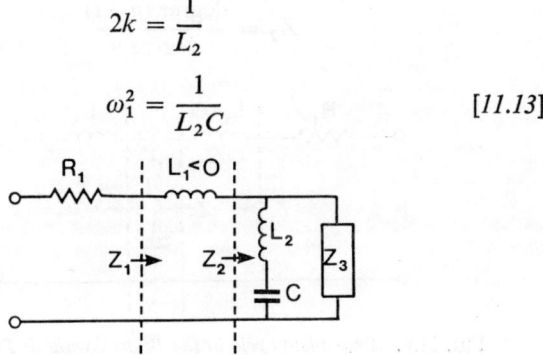

FIG. 11.5. *Second stage in the Brune Synthesis Procedure when $X_1(\omega_1) < 0$.*

Now from [*11.10*] we have

$$Y_3(s) = \frac{1}{Z_1(s) - L_1 s} - \frac{2ks}{s^2 + \omega_1^2}$$

and as $s \to \infty$

$$Y_3(s) \to \frac{1}{-L_1 s} - \frac{2k}{s} = -\left(\frac{1}{L_1} + \frac{1}{L_2}\right)\frac{1}{s}$$

so that Z_3 has a pole at infinity with residue

$$\frac{-L_1 L_2}{L_1 + L_2} = L_3 \qquad [11.14]$$

and since Z_3 is positive real this residue is positive. We next remove this pole at infinity by a series inductor L_3 to give

$$Z_4(s) = Z_3(s) - L_3 s \qquad [11.15]$$

which is positive real. This completes one cycle of the realisation and is shown in Fig. 11.6. Let us now consider the degree of Z_4. Let

$$Z = \frac{\text{degree } n}{\text{degree } n}$$

then

$$Z_1 = \frac{\text{degree } n}{\text{degree } n}$$

$$Z_2 = \frac{\text{degree } (n+1)}{\text{degree } n}$$

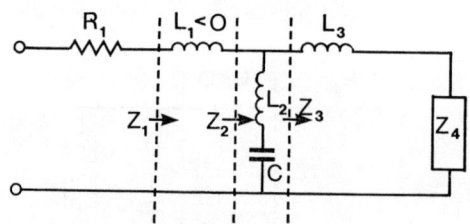

FIG. 11.6. *A complete cycle in the Brune Synthesis Procedure when* $X_1(\omega_1) < 0$.

Now if

$$Y_2 = \frac{A_n(s)}{B_{n+1}(s)}$$

since it has a denominator factor $(s^2 + \omega_1^2)$ we may write

$$Y_2 = \frac{A_n(s)}{(s^2 + \omega_1^2)B'_{n-1}(s)}$$

Y_3 is formed by removing this pole so

$$Y_3 = \frac{A_n(s) - 2ksB'_{n-1}(s)}{(s^2 + \omega_1^2)B'_{n-1}(s)}$$

and the factor $(s^2 + \omega_1^2)$ must cancel in numerator and denominator. Thus

$$Y_3 = \frac{\text{degree } (n-2)}{\text{degree } (n-1)}$$

SYNTHESIS OF RLC DRIVING-POINT IMPEDANCES

Z_3 thus has a pole at infinity and it is removed in forming Z_4. Thus

$$Z_4 = \frac{\text{degree } (n-1)}{\text{degree } (n-2)} - L_3 s$$

$$= \frac{\text{degree } (n-2)}{\text{degree } (n-2)}$$

The cycle has thus produced a positive real remainder, having the same properties as the original but reduced in degree by two. Clearly, repeated application of this process will result in a remainder of degree one or zero which is a resistor or a series circuit, thus

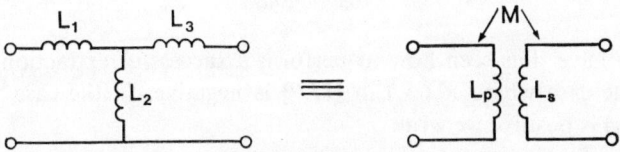

FIG. 11.7. *Equivalence between inductive network and a pair of coupled coils.*

completing the realisation of the original $Z(s)$. We must now show how to remove the negative inductor. This is accomplished by the identity of Fig. 11.7. Calculation of the open-circuit impedance matrix gives

$$L_1 + L_2 = L_p$$
$$L_3 + L_2 = L_s \qquad [11.16]$$
$$L_2 = M$$

From [*11.14*] since the residue is positive, the numerator being positive $L_1 + L_2$ is positive so L_p is positive. L_s is positive since both L_2 and L_3 are positive and M is also positive. Now

$$L_p L_s - M^2 = L_1 L_2 + L_2 L_3 + L_3 L_1$$

But from [*11.14*]

$$L_1 L_2 + L_2 L_3 + L_3 L_1 = 0$$

so that

$$L_p L_s = M^2$$

and the coils are perfectly coupled. Figure 11.6 therefore becomes the circuit shown in Fig. 11.8. Note that the lossless portion of this circuit is the Brune section described in Chapter 10, Section 10.2.1.

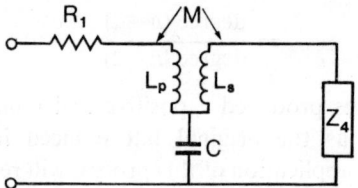

FIG. 11.8. *A complete cycle of the Brune Synthesis Procedure using coupled coils.*

We have thus seen how to perform a successful extraction cycle for the case where $X_1(\omega_1)$ in [*11.9*] is negative. In the case where $X_1(\omega_1)$ is positive we write

$$Y_1(j\omega_1) = jB_1(\omega_1)$$

and $B_1(\omega_1)$ is negative. We then carry out the *dual* of the cycle just described.

$$Y_2(s) = Y_1(s) - C_1 s$$

is formed with C_1 negative and

$$C_1 \omega_1 = B_1(\omega_1)$$

so that $Y_2(s)$ is positive real and is zero at $s = \pm j\omega_1$. $Z_2(s)$ therefore has a pole at these values of s, which is removed by a series circuit leaving a remainder Y_3 with a pole at infinity which is removed by a shunt capacitor to leave a remainder $Z_4(s)$ which is positive real and has the same properties as $Z(s)$ with degree reduced by 2. The resulting circuit is shown in Fig. 11.9.

By duality with [*11.14*] we have

$$C_3 = \frac{-C_1 C_2}{C_1 + C_2} > 0 \qquad [11.17]$$

with $C_2 > 0$, $C_1 < 0$. The problem is then to remove the negative capacitor C_1.

SYNTHESIS OF RLC DRIVING-POINT IMPEDANCES

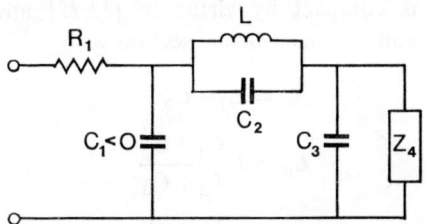

FIG. 11.9. *A complete cycle of the Brune Synthesis Procedure when* $X_1(\omega_1) > 0$.

If we consider only the lossless part of Fig. 11.9 this has an *ABCD* matrix

$$\begin{pmatrix} 1 & 0 \\ C_1 s & 1 \end{pmatrix} \begin{pmatrix} 1 & \dfrac{Ls}{1+LC_2 s^2} \\ 0 & 1 \end{pmatrix} \begin{pmatrix} 1 & 0 \\ C_2 s & 1 \end{pmatrix}$$

$$= \frac{1}{1+LC_2 s^2} \begin{pmatrix} 1+L(C_1+C_2)s^2 & Ls \\ (C_1+C_3)s+L(C_1 C_2+C_2 C_3+C_3 C_1)s^3 & 1+L(C_2+C_3)s^2 \end{pmatrix}$$
[11.18]

But from [11.17] $C_1 C_2 + C_2 C_3 + C_3 C_1 = 0$ and by duality with the case where $X_1(\omega_1) < 0$, C_1+C_2, C_2+C_3 and C_1+C_3 are positive. Thus [11.18] becomes

$$\frac{1}{1+LC_2 s^2} \begin{pmatrix} 1+L(C_1+C_2)s^2 & Ls \\ (C_1+C_3)s & 1+L(C_2+C_3)s^2 \end{pmatrix}$$

so that

$$z_{11} = \frac{1}{(C_1+C_3)s} + L\left(\frac{C_1+C_2}{C_1+C_3}\right)s$$

$$z_{22} = \frac{1}{(C_1+C_3)s} + L\left(\frac{C_2+C_3}{C_1+C_3}\right)s \qquad [11.19]$$

$$z_{12} = \frac{1}{(C_1+C_3)s} + \frac{LC_2}{C_1+C_3}s$$

The network is compact by virtue of [*11.17*] and clearly from Chapter 10 is realisable as a Brune section with

$$C = C_1 + C_3$$

$$L_p = L\frac{C_1+C_2}{C_1+C_3}$$

$$L_s = L\frac{C_2+C_3}{C_1+C_3}$$

$$M = \frac{LC_2}{C_1+C_3} \qquad [11.20]$$

Thus, the realisation of a cycle with $X_1(\omega_1)$ positive is accomplished and completes the proof that the positive real condition is indeed sufficient. Further, the final circuit is the same whether $X_1(\omega_1)$ is positive or negative although it must be derived in a different way for the two cases in order to demonstrate sufficiency. In practical application of the procedure we may apply the simpler procedure for $X_1(\omega_1)$ negative to both cases since although this leads to a non-positive-real impedance when $X_1(\omega_1)$ is positive the proof given above assures us that on completion of a cycle the remainder will still be positive real. The effect of doing this is that in Fig. 11.6 we find $L_1 > 0$ but $L_3 < 0$ and all relationships are still true. A complete realisation according to the Brune method will therefore have the form shown in Fig. 11.10.

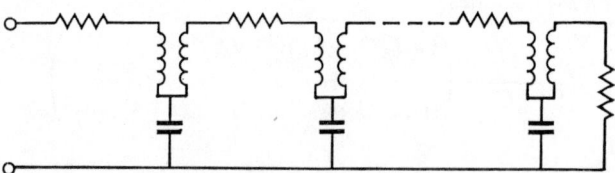

FIG. 11.10. *Realisation of a driving-point impedance by the Brune Procedure.*

As an example of the procedure take

$$Z(s) = 3\frac{2s^2+s+1}{4s^2+2s+1} \qquad [11.21]$$

SYNTHESIS OF RLC DRIVING-POINT IMPEDANCES

The first step is to check for any j-axis poles or zeros and remove them in the Brune preamble. In this example there are no such singularities.

The next step is to calculate the minimum value of $\operatorname{Re} Z(j\omega)$. Now

$$evZ(s) = 3\frac{8s^4+4s^2+1}{16s^4+4s^2+1} \qquad [11.22]$$

so that

$$\operatorname{Re} Z(j\omega) = 3\frac{8\omega^4-4\omega^2+1}{16\omega^4-4\omega^2+1} \qquad [11.23]$$

and we set the derivative of $\operatorname{Re} Z(j\omega)$ to zero to find a minimum. This gives

$$(16\omega^4-4\omega^2+1)(16\omega^2-4) = (8\omega^4-4\omega^2+1)(32\omega^2-4)$$

or

$$32\omega^4-16\omega^2 = 0$$

so that

$$\omega^2 = 0 \text{ or } \tfrac{1}{2}$$

Substitution in [11.23] shows that the minimum occurs at $\omega_1^2 = \tfrac{1}{2}$ and the value is $R_1 = 1$. We then form

$$Z_1(s) = Z(s)-1$$
$$= \frac{2s^2+s+2}{4s^2+2s+1}$$

and $Z_1(j\omega_1) = -j(1/\sqrt{2}) = jX_1$.

We then remove a series inductor

$$L_1 = X_1/\omega_1$$
$$= -1$$

and form

$$Z_2(s) = Z_1(s)-L_1s$$
$$= Z_1(s)+s$$
$$= \frac{4s^3+4s^2+2s+2}{4s^2+2s+1}$$
$$= \frac{4(s^2+\tfrac{1}{2})(s+1)}{4s^2+2s+1}$$

or
$$Y_2(s) = \frac{4s^2+2s+1}{4(s+1)(s^2+\tfrac{1}{2})}$$

Removing the poles at $s = \pm j(1/\sqrt{2})$

$$Y_3(s) = Y_2(s) - \frac{\tfrac{1}{2}s}{(s^2+\tfrac{1}{2})}$$

$$= \frac{1}{2(s+1)}$$

Now
$$Z_3(s) = 2s+2$$

and we remove the pole at infinity by a series inductor $L_3 = 2$ to leave a remainder

$$Z_4 = Z_3 - 2s$$
$$= 2$$

i.e. a 2 ohm resistor. The synthesis is thus complete and the resulting circuit is shown in Fig. 11.11(a). The negative inductor is removed in Fig. 11.11(b) by using the relationships of [11.16].

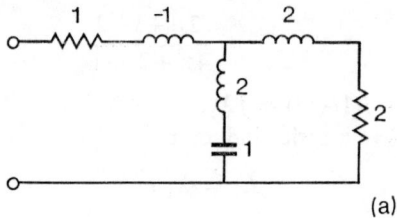

(a)

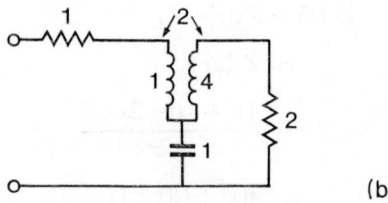

(b)

FIG. 11.11. *Realisation of the impedance function given in [11.21].*

SYNTHESIS OF RLC DRIVING-POINT IMPEDANCES 461

Although the Brune synthesis is quite successful and straightforward to apply the resulting configuration is not very attractive in practice. The appearance of coupled coils is, of course, a complication but in general there is little one can do to avoid them. A more particular disadvantage is that resistors are distributed throughout the network and this, as we shall see, means that the Brune synthesis cannot be adapted for the synthesis of transfer functions. In practice, one is generally much more interested in transfer functions than in driving-point functions but the latter must be studied in detail as the realisation of transfer functions is based upon the synthesis of driving-point impedances.

11.2. Darlington's Synthesis

Although Brune was the first worker to show the sufficiency of the positive real condition for *RLC* driving point impedances, the procedure due to Darlington, first published in 1939, was the step which really made network synthesis the powerful and useful discipline it is today. Darlington, in fact, proved the following theorem.

Darlington's Theorem: Any positive real function can be realised as the driving-point impedance of a *lossless* network terminated in a (one ohm) resistor.

The value of the resistor can be set to one ohm since an ideal transformer incorporated in the lossless network can raise the effective termination to any desired value. The circuit is sketched in Fig. 11.12.

In the Darlington procedure the first step is to carry out the Brune preamble which, of course, yields only lossless elements. The procedure proper therefore begins with a function which is a

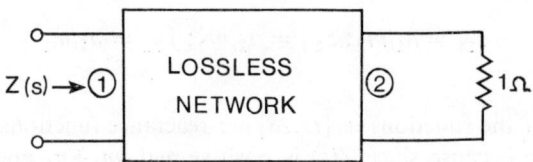

FIG. 11.12. *Realisation of a driving-point impedance by Darlington's Procedure.*

minimum susceptance and a minimum reactance function. Now let the given function be

$$Z(s) = \frac{m_1+n_1}{m_2+n_2} \qquad [11.24]$$

with $m_{1,2}$ even and $n_{1,2}$ odd. Let (z) be the open-circuit impedance matrix of the *lossless* network in Fig. 11.12, so that the input impedance at port 1 is

$$Z_{in} = z_{11} - \frac{z_{12}^2}{1+z_{22}} \qquad [11.25]$$

$$= \frac{z_{11}+|z|}{1+z_{22}}$$

$$= z_{11}\frac{1+|z|/z_{11}}{1+z_{22}}$$

$$= z_{11}\frac{1+1/y_{22}}{1+z_{22}} \qquad [11.26]$$

where y_{22} is the short-circuit driving-point admittance at port 2. We must now try to identify Z_{in} with the prescribed $Z(s)$. To do this we write [11.24] as

$$Z(s) = \frac{m_1}{n_2}\frac{1+(n_1/m_1)}{1+(m_2/n_2)} \qquad \text{Case A}$$

or

$$Z(s) = \frac{n_1}{m_2} = \frac{1+(m_1/n_1)}{1+(n_2/m_2)} \qquad \text{Case B} \qquad [11.27]$$

and equate

$$z_{11} = m_1/n_2;\ z_{22} = m_2/n_2;\ y_{22} = m_1/n_1 \qquad \text{Case A}$$

$$z_{11} = n_1/m_2;\ z_{22} = n_2/m_2;\ y_{22} = n_1/m_1 \qquad \text{Case B}$$

$$[11.28]$$

Now all of the functions in [11.28] are reactance functions. z_{22} and y_{22} are so because since $Z(s)$ is positive real, m_1+n_1 and m_2+n_2 are Hurwitz polynomials, and from Chapter 9 we know that the ratio of the even to the odd part of a Hurwitz polynomial is a reactance function. To see that z_{11} in each case is a reactance function consider the function

$$f(s) = \frac{m_1+n_2}{m_2+n_1}$$

form

$$g(s) = \frac{1-f(s)}{1+f(s)}$$

$$= \frac{m_2-m_1+n_1-n_2}{m_1+m_2+n_1+n_2}$$

Now if $g(s)$ is bounded real $f(s)$ is positive real. Firstly $g(s)$ has no right half plane poles since

$$\Gamma(s) = \frac{1-Z(s)}{1+Z(s)}$$

$$= \frac{m_2-m_1+n_2-n_1}{m_1+m_2+n_1+n_2}$$

is known to be bounded real and has the same denominator. Secondly

$$g(s)g(-s) = \frac{(m_2-m_1)^2+(n_1-n_2)^2}{(m_1+m_2)^2+(n_1+n_2)^2}$$

$$= \Gamma(s)\Gamma(-s)$$

$$\leq 1 \text{ for } s = j\omega$$

Thus $g(s)$ is bounded real and $f(s)$ is positive real so that m_1+n_2 and m_2+n_1 are Hurwitz polynomials. Therefore z_{11} in each case is a reactance function.

Now, although we know that all the functions in [*11.28*] are reactance functions we do not yet know whether they may be realised by a single lossless network. To investigate this we must form z_{12} in each case. This is easily accomplished since

$$1/y_{22} = \frac{|z|}{z_{11}} = z_{22} - \frac{z_{12}^2}{z_{11}}$$

or

$$z_{12} = \left\{z_{11}z_{22} - \frac{z_{11}}{y_{22}}\right\}^{\frac{1}{2}}$$

hence

$$z_{12} = \frac{\sqrt{(m_1 m_2 - n_1 n_2)}}{n_2} \quad \text{Case A}$$

or

$$z_{12} = \frac{\sqrt{(n_1 n_2 - m_1 m_2)}}{m_2} \quad \text{Case B} \qquad [11.29]$$

An immediate difficulty presents itself from [11.29]. It is clearly necessary for any *RLC* network that every network function shall be a rational function of s, whereas the square root in [11.29] will lead to an irrational function for z_{12} unless $m_1 m_2 - n_1 n_2$ is a perfect square. The positive reality of $Z(s)$ does not guarantee in any way that this will be so. To overcome this difficulty we *augment* the given $Z(s)$ by multiplying numerator and denominator by a Hurwitz polynomial $m_0 + n_0$ to give

$$Z(s) = \frac{(m_1 + n_1)(m_0 + n_0)}{(m_2 + n_2)(m_0 + n_0)} \qquad [11.30]$$

$$= \frac{(m_1 m_0 + n_1 n_0) + (m_1 n_0 + n_1 m_0)}{(m_2 m_0 + n_2 n_0) + (m_2 n_0 + n_2 m_0)} \qquad [11.31]$$

$$= \frac{M_1 + N_1}{M_2 + N_2} \qquad [11.32]$$

Notice that although $Z(s)$ is unchanged the even and odd parts of numerator and denominator have changed. Carrying through the same process as before with [11.32] and dealing only with Case A from now on (Case B follows directly) we find

$$z_{11} = \frac{M_1}{N_2}; \ z_{22} = \frac{M_2}{N_2}; \ z_{12} = \frac{\sqrt{(M_1 M_2 - N_1 N_2)}}{N_2} \qquad [11.33]$$

and we must arrange that $M_1 M_2 - N_1 N_2$ is a perfect square by suitable choice of $m_0 + n_0$. Now

$$M_1 M_2 - N_1 N_2 = (m_0^2 - n_0^2)(m_1 m_2 - n_1 n_2) \qquad [11.34]$$

so if we let

$$m_1 m_2 - n_1 n_2 = \prod (s^2 - s_i^2)^{n_i} (s^2 - s_j^2)^{n_j} \qquad [11.35]$$

SYNTHESIS OF RLC DRIVING-POINT IMPEDANCES

where all the n_i are even and all the n_j are odd, then to make [11.34] a perfect square we put

$$m_0^2 - n_0^2 = \prod(s^2 - s_j^2) \qquad [11.36]$$

or

$$(m_0 + n_0)(m_0 - n_0) = \prod(s + s_j)(s - s_j)$$

Now $m_0 + n_0$ must be Hurwitz (to ensure that z_{11} and z_{22} in [11.33] are reactance functions) so we choose all the left half plane factors for $m_0 + n_0$. Thus

$$m_0 + n_0 = \prod(s + s_j) \quad \text{with } \operatorname{Re} s_j > 0. \qquad [11.37]$$

The case where one of the s_j is on the imaginary axis does not arise since all such factors are raised to an even power in $m_1 m_2 - n_1 n_2$. If this were not so then $\operatorname{Re} Z(j\omega)$ would be negative at some point. Thus we can now ensure that z_{12} in [11.33] is rational. It now remains to show that the set of impedance functions in [11.33] satisfies the residue condition $k_{11} k_{22} - k_{12}^2 \geqq 0$ at every pole.

It is clear that poles occur only when $N_2 = 0$ with a possible pole at infinity if the degree of M_1 and M_2 exceeds that of N_1 and N_2. Dealing first with poles occurring when $N_2 = 0$ we have

$$k_{11} = \left\{ \frac{d}{ds}\left(\frac{1}{z_{22}}\right) \right\}^{-1}_{N_2 = 0}$$

$$= \left\{ \frac{M_1 N_2' - N_2 M_1'}{M_1^2} \right\}^{-1}_{N_2 = 0}$$

$$= \left. \frac{M_1}{N_2'} \right|_{N_2 = 0}$$

Similarly

$$k_{22} = \left. \frac{M_2}{N_2'} \right|_{N_2 = 0}$$

and

$$k_{12} = \left\{ \frac{\sqrt{(M_1 M_2 - N_1 N_2 N_2')} - N_2(d/ds)\sqrt{(M_1 M_2 - N_1 N_2)}}{M_1 M_2 - N_1 N_2} \right\}^{-1}_{N_2 = 0}$$

$$= \left. \frac{\sqrt{(M_1 M_2)}}{N_2'} \right|_{N_2 = 0}$$

where the prime denotes differentiation with respect to s. Thus

$$k_{11}k_{22} - k_{12}^2 = 0$$

at every pole corresponding to $N_2 = 0$. In the case of a possible pole at infinity let

$$\frac{M_1}{N_2} \to k_{11}s \quad \text{as} \quad s \to \infty$$

$$\frac{M_2}{N_2} \to k_{22}s \quad \text{as} \quad s \to \infty$$

then

$$z_{12} \to \{k_{11}k_{22}s^2\}^{\frac{1}{2}} = k_{12}s \quad \text{as} \quad s \to \infty$$

since N_1 and N_2 are of the same degree. Again in this case $k_{11}k_{22} - k_{12}^2 = 0$.

Thus at every pole the lossless network satisfies the residue condition with an equality sign so that the network is realisable and compact. Hence the given positive real impedance may always be realised in the form of Fig. 11.12, as stated in Darlington's Theorem.

Darlington's procedure therefore reduces the synthesis of an RLC driving-point impedance to the synthesis of a compact lossless two-port, a problem which has been fully dealt with in Chapter 10. We know from that material that the realisation, in general, requires the use of ideal transformers and/or coupled coils. As an example let us take

$$Z(s) = \frac{2s+2}{s+2} \qquad [11.38]$$

First we check $m_1 m_2 - n_1 n_2$ to see if it is necessary to augment $Z(s)$. In this case

$$m_1 m_2 - n_1 n_2 = 4 - 2s^2$$

which is not a perfect square. We therefore choose $m_0^2 - n_0^2$ to be the product of those factors in $m_1 m_2 - n_1 n_2$ which have odd multiplicity. i.e.

$$m_0^2 - n_0^2 = 4 - 2s^2$$

SYNTHESIS OF RLC DRIVING-POINT IMPEDANCES

and
$$(m_0+n_0)(m_0-n_0) = (2-\sqrt{2}s)(2+\sqrt{2}s)$$
so that
$$m_0+n_0 = 2+\sqrt{2}s$$
Therefore we write
$$Z(s) = \frac{(2s+2)(2+\sqrt{2}s)}{(s+2)(2+\sqrt{2}s)}$$
and
$$M_1 = 4+2\sqrt{2}s^2, \quad M_2 = 4+\sqrt{2}s^2$$
$$N_1 = (4+2\sqrt{2})s, \quad N_2 = (2+2\sqrt{2})s$$

These give for the lossless network
$$z_{11} = \frac{M_1}{N_2} = \frac{2+\sqrt{2}s^2}{(1+\sqrt{2})s}$$
$$z_{22} = \frac{M_2}{N_2} = \frac{4+\sqrt{2}s^2}{(2+2\sqrt{2})s} \qquad [11.39]$$
$$z_{12} = \frac{2-s^2}{(1+\sqrt{2})s}$$

[11.39] must now be realised as a lossless network. Rewriting we obtain
$$z_{11} = \frac{2}{(1+\sqrt{2})s} + \frac{\sqrt{2}}{1+\sqrt{2}}s$$
$$z_{22} = \frac{2}{(1+\sqrt{2})s} + \frac{1}{\sqrt{2}(1+\sqrt{2})}s$$
$$z_{12} = \frac{2}{(1+\sqrt{2})s} - \frac{1}{1+\sqrt{2}}s$$

This is clearly a series connection of two networks as shown in Fig. 11.13(a) in which
$$C = \frac{1+\sqrt{2}}{2}$$
$$L = \frac{\sqrt{2}}{1+\sqrt{2}}$$
$$n = -\frac{1}{\sqrt{2}}$$

G

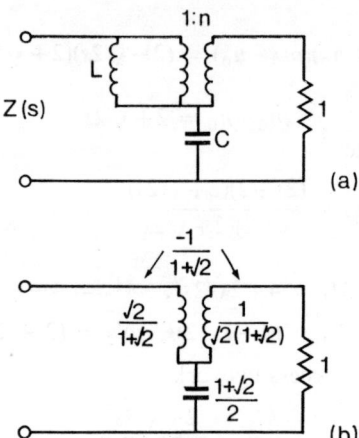

Fig. 11.13. *Realisation of the impedance given in [11.38]*. (a) *General form of the circuit;* (b) *The specific circuit, using coupled coils.*

and converting to coupled coils we obtain the network of Fig. 11.13(*b*). Note that M is negative and the lossless part of the network corresponds to the Darlington C-section described in Chapter 10.

11.3. Bott and Duffin Synthesis

We have seen already that a disadvantage of the Darlington and Brune procedures is the appearance, in general, of coupled coils. However, it is always possible to realise any positive real function as the driving point impedance of a network without ideal transformers or coupled coils. This was first shown by Bott and Duffin. Although the network resulting from their procedure does not contain ideal transformers or coupled coils it does require a relatively large number of elements and it is rather sensitive to the values of these elements. The procedure also requires the resistance to be distributed through the network and is therefore of no use in synthesising transfer functions. The main interest is therefore theoretical, and in that respect the method is of major importance in demonstrating that, in general, ideal transformers or coupled coils are not necessary.

The procedure is based on a result known as Richards' Theorem (a result which is of considerable importance in the area of distri-

SYNTHESIS OF RLC DRIVING-POINT IMPEDANCES   469

buted network theory). The theorem states that if $Z(s)$ is positive real then

$$Z_1(s) = \frac{kZ(s)-sZ(k)}{kZ(k)-sZ(s)} \qquad [11.40]$$

is also positive real for any real positive value of k. $Z(k)$ is the value of $Z(s)$ obtained by replacing s by k, and is real and positive since $Z(s)$ is positive real.

To show that $Z_1(s)$ is positive real we form the associated reflection coefficient

$$\Gamma_1(s) = \frac{1-Z_1(s)}{1+Z_1(s)} = \frac{(k+s)\{Z(k)-Z(s)\}}{(k-s)\{Z(k)+Z(s)\}}$$

$$= \frac{k+s}{k-s}\left\{\frac{1-Z(s)/Z(k)}{1+Z(s)/Z(k)}\right\} \qquad [11.41]$$

To show that $Z_1(s)$ is positive real we need only show that $\Gamma_1(s)$ is bounded real (Chapter 9, Section 9.3). The bracketed portion of [11.41] is bounded real since $Z(s)$ (and hence $Z(s)/Z(k)$) is positive real. Thus we need only to show that $\Gamma_1(s)$ has no right half plane poles and $|\Gamma_1(j\omega)| \leq 1$. Writing [11.41] as

$$\Gamma_1(s) = \frac{k+s}{k-s}\Gamma(s) \qquad [11.42]$$

we find that

$$|\Gamma_1(j\omega)|^2 = \frac{k^2+\omega^2}{k^2+\omega^2}|\Gamma(j\omega)|^2$$

But $|\Gamma(j\omega)| \leq 1$ so therefore $|\Gamma_1(j\omega)| \leq 1$. From [11.42] it appears that $\Gamma_1(s)$ has a right half plane pole at $s = k$ but from [11.41] $\Gamma(s)$ has a zero at this point so that the factor $k-s$ cancels. Also since $\Gamma(s)$ is bounded real it cannot have any right half plane poles so that $\Gamma_1(s)$ has no right half plane poles and is therefore bounded real. Hence $Z_1(s)$ is positive real.

$Z_1(s)$ as given in [11.40] is not of higher degree than $Z(s)$ since both numerator and denominator have a factor $s-k$ which therefore cancels.

The Bott and Duffin procedure begins by carrying out the Brune preamble and then removing the minimum value of the real part of the impedance, just as in [11.8] of the Brune synthesis. We are thus concerned with an impedance $Z(s)$, such that it has no j-axis poles or zeros and

$$Z(j\omega_1) = jX_1(\omega_1) \qquad [11.43]$$

Again we have two possibilities ($X_1 > 0$ and $X_1 < 0$) which must be treated separately. We shall take the case of $X_1 > 0$. Now let

$$Z(j\omega_1) = jX_1(\omega_1) = jL_1\omega_1 \qquad [11.44]$$

and we shall choose k in [11.40] such that

$$L_1 = Z(k)/k > 0$$

Note that this implies that the numerator of [11.40] has a zero at $s = j\omega_1$ since it reads $k(Z(j\omega_1) - j\omega_1 L_1)$ at that frequency. Rearranging [11.40] we obtain

$$\begin{aligned} Z(s) &= \frac{Z(k)[kZ_1(s)+s]}{k+sZ_1(s)} \\ &= \frac{kZ(k)Z_1(s)}{k+sZ_1(s)} + \frac{sZ(k)}{k+sZ_1(s)} \\ &= \frac{1}{[1/Z(k)Z_1(s)]+[s/kZ(k)]} + \frac{1}{[k/sZ(k)]+[Z_1(s)/Z(k)]} \end{aligned}$$
$$[11.45]$$

This is interpreted in circuit terms as shown in Fig. 11.14 in which

$$C_B = \frac{1}{kZ(k)}$$

$$Z_B = Z(k)Z_1(s)$$

$$L_A = \frac{Z(k)}{k} = L_1$$

$$Z_A = \frac{Z(k)}{Z_1(s)} \qquad [11.46]$$

Since, by Richards' theorem $Z_1(s)$ is positive real, Z_A and Z_B are positive real, and are of the same degree as $Z(s)$. L_A and C_B are positive. Now since $Z_1(s)$ has been constructed to have zeros at $s = \pm j\omega_1$, Z_A has corresponding poles and Z_B corresponding

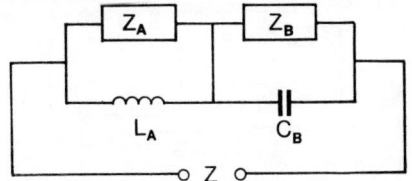

FIG. 11.14. *Realisation of an impedance by the Bott–Duffin method according to [11.45] with $X_1(\omega_1) > 0$.*

zeros. Since both are positive real these may be extracted as shown in Fig. 11.15 to leave positive real remainders

$$Z'_A = Z_A - \frac{2k_1 s}{s^2 + \omega_1^2} \qquad [11.47]$$

$$Y'_B = Y_B - \frac{2\alpha_1 s}{s^2 + \omega_1^2} \qquad [11.48]$$

whose degrees are two less than the original $Z(s)$. The procedure may then be repeated on Z'_A and Z'_B and so on until the entire network is realised. It is clear from the description given above that a large number of elements are needed but that the use of transformers or coils is avoided. If the circuit of Fig. 11.14 is redrawn as in Fig. 11.16 it is clear that the realisation is in the form of a balanced bridge since from [11.46]

$$Z_A Z_B = L_A/C_B$$

Thus the short-circuit between A and B may be removed. This observation explains the sensitivity of the performance to the values

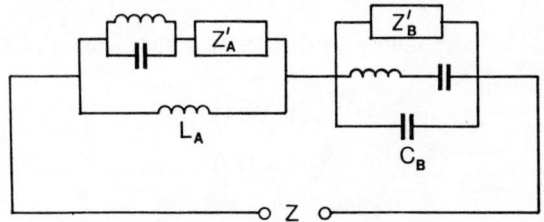

FIG. 11.15. *A complete cycle of the Bott–Duffin Synthesis Procedure.*

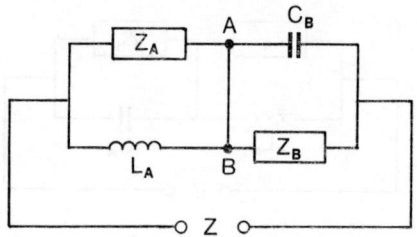

FIG. 11.16. *A rearranged version of Fig. 11.14 to show that it is a balanced bridge.*

of the elements and has led to several variations on the basic Bott–Duffin technique (see for example E. A. Guillemin, 1957. *Synthesis of Passive Networks*, J. Wiley & Sons, New York and London, for further details).

In the case where $X_1(\omega_1)$ in [*11.44*] is negative we must switch to an admittance basis by replacing Z by Y in [*11.40*] and $Y_1(s)$ is then positive real. This gives

$$Y(j\omega_1) = jB_1(\omega_1) = jC_1\omega_1$$

with $B_1 > 0$. We then put

$$\frac{Y(k)}{k} = C_1$$

$$Y(s) = \frac{Y(k)[kY_1(s)+s]}{k+sY_1(s)}$$

$$= \frac{1}{[1/Y(k)Y_1(s)]+[s/kY(k)]} + \frac{1}{[k/sY(k)]+[Y_1(s)/Y(k)]}$$

[*11.49*]

The circuit corresponding to this form is that shown in Fig. 11.17 in which

$$Z^{(1)} = 1/Y(k)Y_1(s)$$
$$L^{(1)} = 1/kY(k)$$
$$C^{(2)} = Y(k)/k = C_1$$
$$Z^{(2)} = Y_1(s)/Y(k)$$

[*11.50*]

SYNTHESIS OF RLC DRIVING-POINT IMPEDANCES

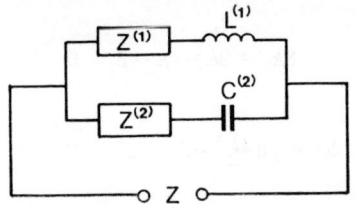

FIG. 11.17. *Realisation of an impedance by the Bott–Duffin method in the case where* $X_1(\omega_1) < 0$.

and we note that this is the same as the circuit of Fig. 11.16 with the short-circuit removed. Thus both cases may be realised by a circuit of the same form. Since $Y_1(s)$ is zero at $s = \pm j\omega_1$ the cycle is completed by removing the poles at $s = \pm j\omega_1$ from $Z^{(1)}$ and $1/Z^{(2)}$.

As an example of the Bott and Duffin procedure consider the impedance function given in [*11.21*]

$$\hat{Z}(s) = 3\,\frac{2s^2+s+1}{4s^2+2s+1} \qquad [11.51]$$

as before we remove a resistor equal to the minimum value of $\mathrm{Re}\hat{Z}(j\omega)$ to leave a remainder

$$Z(s) = \frac{2s^2+s+1}{4s^2+2s+1} \qquad [11.52]$$

with $\mathrm{Im}Z(j\omega_1) = -j/\sqrt{2}$ at $\omega_1 = 1/\sqrt{2}$.

Now since $X_1 < 0$ we must switch to an admittance basis to give

$$Y(j\omega_1) = j\sqrt{2} = jC_1\omega_1$$

or

$$C_1 = 2$$

Now we put

$$\frac{Y(k)}{k} = 2$$

or

$$\frac{2k^2+k+2}{4k^2+2k+1} = 2k$$

giving
$$8k^3 + 2k^2 + k - 2 = 0$$
or
$$2(k - \tfrac{1}{2})(4k^2 + 3k + 2) = 0$$
i.e.
$$k = \tfrac{1}{2}$$

Now from [11.50]
$$Z^{(1)} = 1/Y_1(s)$$
$$L^{(1)} = 2$$
$$C^{(2)} = 2$$
$$Z^{(2)} = Y_1(s)$$

where
$$Y_1(s) = \frac{2s^2 + 1}{4s^2 + 3s + 2}$$

when a factor $s - \tfrac{1}{2}$ is cancelled in numerator and denominator. Now

$$\frac{1}{Y_1(s)} = 2 + \frac{3s}{2s^2 + 1} = Z^{(1)} = \frac{1}{Z^{(2)}}$$

and the realisation is that shown in Fig. 11.18.

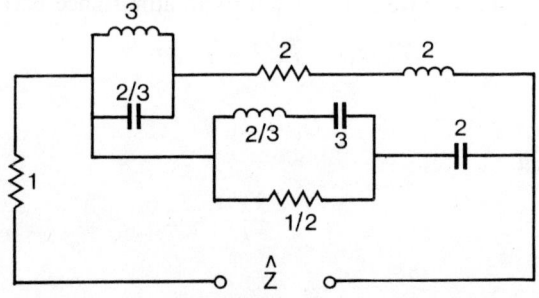

FIG. 11.18. *Realisation of the impedance function given in* [11.51] *by the Bott–Duffin Procedure.*

11.4. Conclusions

In this chapter we have seen the classical realisation techniques of Brune, Darlington, and Bott and Duffin, for a general positive real driving-point impedance. Except for the latter case, ideal transformers and/or coupled coils are normally required. We shall see in the next chapter that the Darlington driving point impedance synthesis procedure forms the basis for transfer function synthesis and it is from this fact that its major importance derives.

Problems

11.1. Find the conditions on γ such that the function given below is realisable by the circuit of Fig. 11.1 and synthesise the function

$$Z(s) = \frac{s+\gamma}{s^2+4s+8}$$

11.2. Synthesise the following impedance functions using the Brune Procedure.

(a) $\dfrac{18s^2+4s+3}{8s^2+s+1}$

(b) $3\dfrac{3s^2+s+1}{s^2+2s+1}$

(c) $\dfrac{8s^3+19s^2+5s+3}{8s^2+s+1}$

(d) $\dfrac{18s^4+4s^3+21s^2+4s+3}{8s^4+19s^3+13s^2+4s+1}$

11.3. Synthesise the following impedance functions using the Darlington Procedure.

(a) $Z(s) = 4\dfrac{s^2+s+1}{s^2+s+4}$

(b) $Z(s) = \dfrac{s^2+4s+1}{s^2+s+1}$

(c) $Z(s) = \dfrac{2s+1}{8s+1}$

(d) $Z(s) = \dfrac{s^4+4s^3+3s^2+4s+1}{s^4+s^3+3s^2+s+1}$ ($m_1m_2-n_1n_2$ is a perfect square)

(e) $Z(s) = \dfrac{2s^3+3s^2+6s+1}{s^2+s+1}$

(f) $Z(s) = \dfrac{2s^3+3s^2+2s+1}{s^2+s+1}$ *without* ideal transformers *or* coupled coils.

(g) $Z(s) = \dfrac{2s^2+2s+1}{2s^3+2s^2+2s+1}$

11.4. Synthesise the impedance functions given in Problems 11.2(*a*) and (*b*) using the Bott–Duffin Technique.

Chapter 12

TRANSFER FUNCTION SYNTHESIS

In this chapter we shall deal with the synthesis of transfer functions for resistively-terminated lossless networks. This type of network corresponds to the practical requirements of signal processing where it is desired to operate upon a signal without wasteful dissipation of power in the network. We have already seen, in Chapter 10, how to synthesise transfer functions of purely lossless networks, but in practice it is normally found that either the signal source or the load, or both, have resistive impedances which cannot be ignored relative to the impedance level of the remainder of the network so that resistively terminated lossless networks are representative of most practical situations. A double-terminated network is shown in

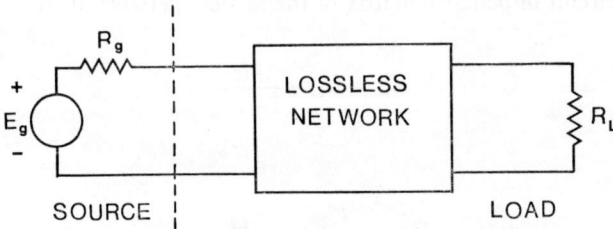

FIG. 12.1. *Double-terminated lossless network.*

Fig. 12.1 while if the ratio of load to source resistance is either very high or very low it may be possible to neglect one or other resistor relative to the impedances in the remainder of the network, leading to what is effectively a single-terminated structure, as shown in Figs. 12.2(*a*) and (*b*). Different synthesis techniques are required for single and double-terminated networks.

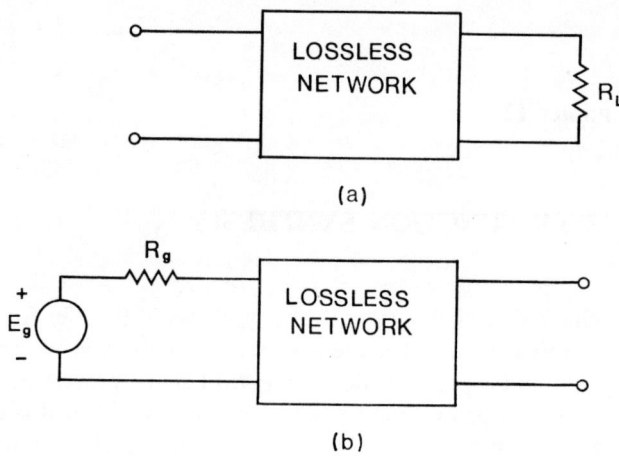

FIG. 12.2. *Single-terminated lossless networks* (a) *when* $R_L \gg R_g$; (b) *when* $R_g \gg R_L$.

12.1. Some General Properties

We begin by developing some general properties of lossless terminated networks. Consider the network of Fig. 12.3. If (z) is the open-circuit impedance matrix of the lossless network then

$$Z_{in} = z_{11} - \frac{z_{12}^2}{1+z_{22}} \qquad [12.1]$$

and

$$Z_{21} = \frac{V_2}{I_1} = \frac{z_{12}}{1+z_{22}} \qquad [12.2]$$

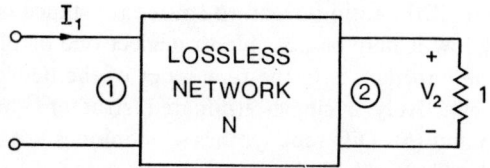

FIG. 12.3. *Current-driven single-terminated lossless network.*

TRANSFER FUNCTION SYNTHESIS

Now if we form

$$Z_{in}(s)+Z_{in}(-s) = z_{11}(s)-\frac{z_{12}^2(s)}{1+z_{22}(s)}+z_{11}(-s)-\frac{z_{12}^2(-s)}{1+z_{22}(-s)}$$

then since (z) refers to a lossless network z_{11}, z_{22} and z_{12} are odd functions so that

$$z_{11}(-s) = -z_{11}(s)$$
$$z_{22}(-s) = -z_{22}(s)$$
$$z_{12}(-s) = -z_{12}(s)$$

Hence

$$Z_{in}(s)+Z_{in}(-s) = \frac{-2z_{12}^2(s)}{1-z_{22}^2(s)} \qquad [12.3]$$

also

$$Z_{21}(s)Z_{21}(-s) = \frac{-z_{12}^2(s)}{1-z_{22}^2(s)} \qquad [12.4]$$

$$= \tfrac{1}{2}[Z_{in}(s)+Z_{in}(-s)] \qquad [12.5]$$

[12.5] expresses the fact that since the network is lossless all the power entering at port 1 is dissipated in the load. The power in the load is

$$P_L = V_2(j\omega)V_2(-j\omega) \qquad [12.6]$$

while the power entering at port 1 is

$$P_{in} = I_1(j\omega)I_1(-j\omega)\,\text{Re}\,Z_{in}(j\omega) \qquad [12.7]$$

Equating these two quantities gives

$$Z_{21}(j\omega)Z_{21}(-j\omega) = \text{Re}\,Z_{in}(j\omega)$$
$$= \tfrac{1}{2}[Z_{in}(j\omega)+Z_{in}(-j\omega)] \qquad [12.8]$$

and replacing $j\omega$ by s leads to [12.5]. The zeros of $Z_{21}(s)Z_{21}(-s)$ are the values of s for which no power appears in the load (the definition of power being extended from real frequencies to complex frequencies), and are therefore properly called the zeros of transmission. [12.5] means that for the circuit of Fig. 12.3 the zeros of the even part of $Z_{in}(s)$ are identical with the zeros of transmission, including multiplicity. It is important to note that this is true only because N is a *lossless* network. We therefore often class a driving-point impedance in terms of its 'zeros of transmission' by which is meant the zeros of its even part.

If the network is driven by a source of finite internal resistance as shown in Fig. 12.4 then the input power (equal to the power in the load) is

$$P_{in} = \tfrac{1}{2}I_1(j\omega)I_1(-j\omega)[Z_{in}(j\omega)+Z_{in}(-j\omega)]$$

$$= \tfrac{1}{2}\frac{E_g}{R_g+Z_{in}(j\omega)} \cdot \frac{E_g^*}{R_g+Z_{in}(-j\omega)}[Z_{in}(j\omega)+Z_{in}(-j\omega)]$$

[12.9]

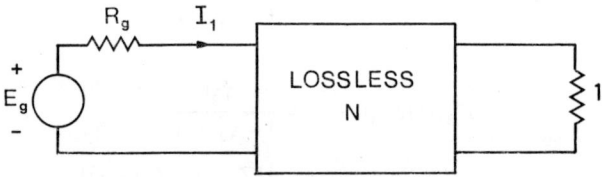

FIG. 12.4. *Double-terminated lossless network.*

or replacing $j\omega$ by s, the zeros of transmission are the zeros of

$$\frac{Z_{in}(s)+Z_{in}(-s)}{\{R_g+Z_{in}(s)\}\{R_g+Z_{in}(-s)\}} \qquad [12.10]$$

Thus additional transmission zeros (apart from the zeros of ev $Z(s)$) occur at values of s which are simultaneously poles of $Z_{in}(s)$ and $Z_{in}(-s)$. But poles of both $Z_{in}(s)$ and $Z_{in}(-s)$ can occur only on the imaginary axis so that the only possible exceptions are zeros of transmission on the imaginary axis which are simultaneously poles of $Z_{in}(s)$ or by duality poles of $Y_{in}(s)$.

From [12.9] the transducer power gain (ratio of power in the load to available power from the generator) is

$$|S_{12}(j\omega)|^2 = \frac{2[Z_{in}(j\omega)+Z_{in}(-j\omega)]R_g}{\{R_g+Z_{in}(j\omega)\}\{R_g+Z_{in}(-j\omega)\}} \qquad [12.11]$$

12.2. Synthesis of Single-terminated Networks

In this section we shall consider the synthesis of networks of the type shown in Fig. 12.3, having a transfer impedance as given by [*12.2*]. Firstly, it is necessary that the specified $Z_{21}(s)$ have a numerator which is an even or odd function of s (since it is also the numerator of z_{12}), and, secondly, the denominator must be a Hurwitz polynomial since it is the sum of the numerator and denominator of z_{22}. Thus if

$$Z_{21} = \frac{E}{m+n} \quad \text{Case A}$$

or

$$Z_{21} = \frac{O}{m+n} \quad \text{Case B}$$

[*12.12*]

with m even and n odd we may immediately divide to obtain

$$Z_{21} = \frac{E/n}{1+(m/n)} \quad \text{Case A}$$

or

$$Z_{21} = \frac{O/m}{1+(n/m)} \quad \text{Case B}$$

[*12.13*]

and from [*12.2*]

$$z_{12} = (E/n); \; z_{22} = (m/n) \quad \text{Case A}$$
$$z_{12} = (O/m); \; z_{22} = (n/m) \quad \text{Case B}$$

[*12.14*]

which are associated with the lossless network. The synthesis of the lossless network could be completed by assigning suitable functions for z_{11} so that $k_{11}k_{22} - k_{12}^2 \geqq 0$ at every pole and the techniques of Section 10.2.1 could be used to realise the network. Thus success is always assured provided the numerator of Z_{21} is even or odd, and the denominator is Hurwitz. As an example consider

$$Z_{21} = \frac{2s^2+1}{4s^2+s+1} \quad [12.15]$$

The numerator is even, so we have Case A with $m = 4s^2+1; n = s$. Thus

$$z_{12} = \frac{2s^2+1}{s}; z_{22} = \frac{4s^2+1}{s} \quad [12.16]$$

We can now choose any appropriate z_{11} but it is obviously convenient to make the choice so that the network is compact. We therefore choose

$$z_{11} = s+(1/s)$$

The z-matrix of the lossless network is then the same as that used in the example given in [10.37] and the realisation of Z_{21} is shown in Fig. 12.5.

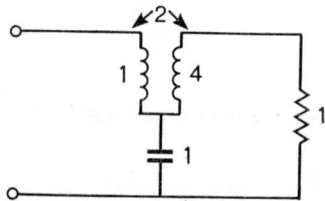

Fig. 12.5. *Realisation of the impedance given in* [12.15].

The realisation procedure given above, while always guaranteed to be successful, suffers from the disadvantage that coupled coils and/or ideal transformers are usually required as we have already seen in Chapter 10. This may be avoided in many cases since the specification of Z_{21} does not involve the value of z_{11}, while the procedure outlined above involves assigning a value to this function. The basic problem, however, only calls for the simultaneous realisation of *two* prescribed impedance functions for the lossless network. This is similar to the problem considered in Section 10.5, but in the present case an *exact* realisation is required. This is always possible when the zeros of transmission (zeros of Z_{21} and z_{12}) lie only at the origin or infinity. We then merely synthesise z_{22} in the form of a ladder with series inductance and shunt capacitance or *vice versa*. For example if

$$Z_{21} = \frac{1}{s^3+2s^2+2s+1} \quad [12.17]$$

we have case A, and

$$z_{12} = \frac{1}{s^3+2s}; \; z_{22} = \frac{2s^2+1}{s^3+2s} \qquad [12.18]$$

and to get the correct zeros of transmission we need series inductors and shunt capacitors. We therefore write

$$\frac{1}{z_{22}} = \frac{s^3+3s}{2s^2+1}$$

and expand in a continued fraction

$$\frac{1}{z_{22}} = 2s^2+1 \overline{\smash{\big)}\, s^3+2s} \quad \begin{array}{c} s/2 \\ \end{array}$$

$$\underline{s^3 + \frac{s}{2}} \qquad 4s/3$$

$$\frac{3}{2}s \overline{\smash{\big)}\, 2s^2+1}$$

$$\underline{2s^2} \qquad 3s/2$$

$$1 \overline{\smash{\big)}\, \frac{3}{2}s}$$

$$\underline{\frac{3}{2}s}$$

$$0$$

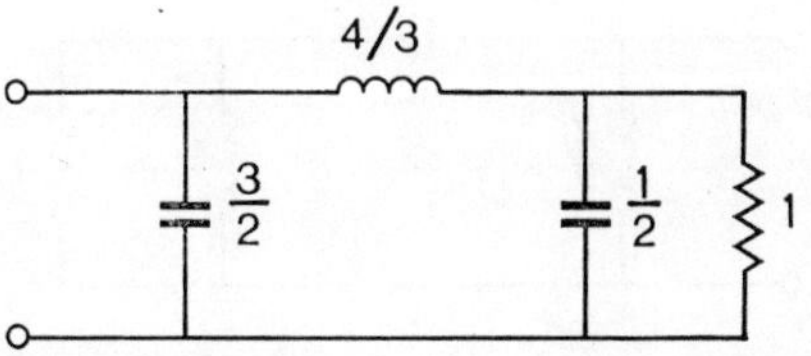

FIG. 12.6. *Realisation of the open-circuit transfer impedance given in* [*12.17*].

The resulting network is that shown in Fig. 12.6. Note that $Z_{21}(0) = 1$. To realise KZ_{12} we merely impedance-scale the network by a factor K resulting in a terminating resistor of K ohms.

An example with transmission zeros at the origin is

$$Z_{21} = \frac{s^3}{s^3+2s^2+2s+1} \qquad [12.19]$$

giving case B with

$$z_{12} = \frac{s^3}{2s^2+1}; \; z_{22} = \frac{s^3+2s}{2s^2+1} \qquad [12.20]$$

We require shunt inductance and series capacitance so we write

$$\frac{1}{z_{22}} = \frac{1+2s^2}{2s+s^3}$$

and expand in a continued fraction to give

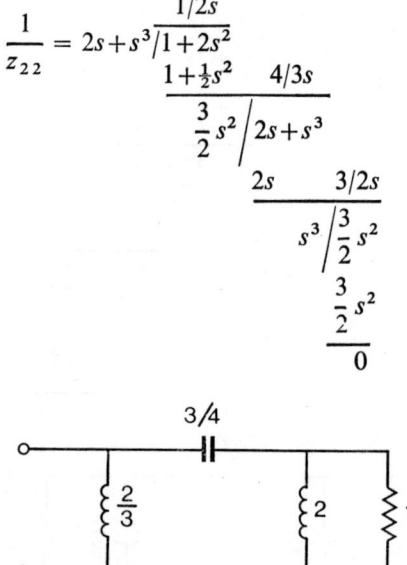

FIG. 12.7. *Realisation of the open-circuit transfer impedance given in* [12.19].

and the circuit of Fig. 12.7. In this case $Z_{21}(\infty) = 1$, and KZ_{21} may be realised by impedance scaling.

Zeros of transmission on the imaginary axis can be realised by a zero-shifting procedure of the type described in Section 10.5.2, since we always use impedance scaling to get the exact value of Z_{21}. Guillemin's procedure as described in Section 10.5.3 may also be used in this context.

12.2.1. REALISATION OF $|Z_{21}(j\omega)|$

We have just seen how to realise a specified $Z_{21}(s)$ by a single terminated lossless network. In practice, however, we are normally only interested in the behaviour of this function at real frequencies i.e. the power in the load, and so it is usually $|Z_{21}(j\omega)|^2$ which is specified rather than $Z_{21}(s)$. Since the numerator of $Z_{21}(s)$ is an even or odd function, the numerator of $|Z_{21}(j\omega)|^2$ must be the square of an even or odd function. Thus

$$|Z_{21}(j\omega)|^2 = \frac{A^2(\omega)}{B(\omega^2)} \qquad [12.21]$$

where $A(\omega)$ is even or odd. We must now form $Z_{21}(s)$ from [12.21]. From [12.12]

$$Z_{21}(j\omega)Z_{21}(-j\omega) = \left.\frac{E^2(s) \text{ or } O^2(s)}{m^2(s)-n^2(s)}\right|_{s=j\omega} \qquad [12.22]$$

Replacing ω^2 by $-s^2$ in [12.21] gives

$$Z_{21}(s)Z_{21}(-s) = \frac{A^2(s)}{B(-s^2)} \qquad [12.23]$$

Thus

$$E(s) \text{ or } O(s) = A(s) \qquad [12.24]$$

and

$$m^2(s)-n^2(s) = B(-s^2) \qquad [12.25]$$

We must now determine $m(s)+n(s)$. We know that this function must be Hurwitz, i.e. must have only left-half-plane zeros so we factorise $B(-s^2)$ and assign the factors with left-half-plane zeros to $m+n$ and those with right-half-plane zeros to $m-n$. As an example take

$$|Z_{21}(j\omega)|^2 = \frac{1}{1+\omega^6}$$

$$Z_{21}(s)Z_{21}(-s) = \frac{1}{1-s^6} \qquad [12.26]$$

so that

$$E(s) = 1$$
$$m^2-n^2 = 1-s^6 = (1-s)(1+s)(s^2+s+1)(s^2-s+1)$$

486 CIRCUIT THEORY

We therefore take the factors with left-half-plane roots for $m+n$ to give

$$m+n = (s+1)(s^2+s+1)$$
$$= s^3+2s^2+2s+1$$

and

$$Z_{21}(s) = \frac{1}{s^3+2s^2+2s+1} \quad [12.27]$$

which is the example considered in [12.17] leading to the circuit of Fig. 12.6 which therefore has $|Z_{21}(j\omega)|^2$ as given in [12.26]. This is a low-pass filter characteristic of the maximally flat class which will be described later in this chapter.

12.3. Double-loaded Networks

We are interested, in this section, in the transducer power gain of a network of the type shown in Fig. 12.4. This is given by [12.11] as

$$|S_{12}(j\omega)|^2 = \frac{2[Z_{in}(j\omega)+Z_{in}(-j\omega)]R_g}{[R_g+Z_{in}(j\omega)][R_g+Z_{in}(-j\omega)]} \quad [12.28]$$

so that

$$1-|S_{12}(j\omega)|^2 = \frac{\{\pm R_g \mp Z_{in}(j\omega)\}\{\pm R_g \mp Z_{in}(-j\omega)\}}{\{R_g+Z_{in}(j\omega)\}\{R_g+Z_{in}(-j\omega)\}} \quad [12.29]$$

$$= \left|\frac{\pm R_g \mp Z_{in}(j\omega)}{R_g+Z_{in}(j\omega)}\right|^2 \quad [12.30]$$

But, according to [9.56] since, N is lossless

$$1-|S_{12}(j\omega)|^2 = |S_{11}(j\omega)|^2 \quad [12.31]$$

and from [9.54]

$$\frac{Z_{in}}{R_g} = \frac{1+S_{11}}{1-S_{11}} \quad [12.32]$$

and

$$S_{11}(j\omega) = \frac{Z_{in}(j\omega)-R_g}{Z_{in}(j\omega)+R_g} \quad [12.33]$$

The synthesis problem to be solved is the realisation of a specified transducer power gain by a double-terminated lossless network. Thus $|S_{12}(j\omega)|^2$ is specified and the procedure is

(i) Form $|S_{11}(j\omega)|^2$ using [*12.31*]

(ii) Form $S_{11}(p)S_{11}(-p)$ by replacing $j\omega$ by p

(iii) Form $S_{11}(p)$ as described below

(iv) Form $Z_{in}(p)/R_g$ from [*12.32*]

(v) Synthesise $Z_{in}(p)$ by Darlington's procedure

(where p is used in place of s to avoid confusion).

To form $S_{11}(p)$ from $S_{11}(p)S_{11}(-p)$ we note that from [*12.33*] $S_{11}(p)$ has no right-half-plane poles (if it had, Z_{in} would equal $-R_g$ at such a pole, in contradiction to the requirement that it be positive real). Thus we factorise the denominator of $S_{11}(p)S_{11}(-p)$ and assign those factors with left-half-plane roots to $S_{11}(p)$. No corresponding restriction on the numerator exists since $S_{11}(p)$ can have right-half-plane zeros when Z_{in} is positive real (Z_{in} is equal to R_g at such zeros). Thus if we factorise the numerator of $S_{11}(p)S_{11}(-p)$ we may assign the factors in any way between $S_{11}(p)$ and $S_{11}(-p)$. Note that this, in general, yields a number of different functions for Z_{in} and consequently gives a number of different networks, all of which yield the *same* function $|S_{12}(j\omega)|^2$. It is often convenient to choose the zeros of $S_{11}(p)$ to be in the left-half-plane, although in a practical situation various possibilities may be tried in order to arrive at the most desirable network.

The procedure given above guarantees that Z_{in} is positive real provided that $0 \leq |S_{12}(j\omega)|^2 \leq 1$ since then $0 \leq |S_{11}|^2 \leq 1$ and we choose the denominator of $S_{11}(p)$ to avoid right-half-plane poles. Thus $S_{11}(p)$ is bounded-real which ensures that $Z_{in}(p)$ is positive real according to section 9.5.

As an example of the above procedure consider the transducer power gain

$$|S_{12}(j\omega)|^2 = \frac{1}{1+\omega^6}$$

The procedure then gives

(i) $|S_{11}(j\omega)|^2 = 1 - |S_{12}(j\omega)|^2$

$$= \frac{\omega^6}{1+\omega^6}$$

(ii) $S_{11}(p)S_{11}(-p) = \dfrac{-p^6}{1-p^6}$

(iii) The denominator is easily factorised as

$$1-p^6 = (1-p)(1+p)(p^2+p+1)(p^2-p+1)$$

and we choose those factors with left-half-plane roots for $S_{11}(p)$. The zeros of the numerator all lie at the origin, so that no choice is possible here. Thus

$$S_{11}(p) = \frac{\pm p^3}{(1+p)(p^2+p+1)}$$

$$= \frac{\pm p^3}{p^3+2p^2+2p+1}$$

(iv) $\dfrac{Z_{in}}{R_g} = \dfrac{1+S_{11}}{1-S_{11}}$

$$= \frac{(p^3 \pm p^3)+2p^2+2p+1}{(p^3 \mp p^3)+2p^2+2p+1} \quad [12.34]$$

We then realise Z_{in} by Darlington's procedure.

12.3.1. LADDER NETWORKS

The procedure of the previous section requires the realisation of a general positive real driving point impedance as the final step. In many important cases, however, the prescribed transducer power gain is of the form

$$|S_{12}(j\omega)|^2 = \frac{1}{D_n(j\omega)D_n(-j\omega)}$$

so that all the zeros of transmission are at infinity i.e. $|S_{12}| = 0$ only at $\omega = \infty$. In the synthesis procedure we form

$$|S_{11}(j\omega)|^2 = 1 - |S_{12}(j\omega)|^2$$

$$= \frac{D_n(j\omega)D_n(-j\omega) - 1}{D_n(j\omega)D_n(-j\omega)}$$

$$S_{11}(p)S_{11}(-p) = \frac{D_n(p)D_n(-p) - 1}{D_n(p)D_n(-p)}$$

Now as $p \to \infty$

$$S_{11}(p)S_{11}(-p) \to 1$$

so

$$S_{11}(p) \to \pm 1$$

But

$$S_{11}(p) = \frac{\dfrac{Z_{in}(p)}{R_g} - 1}{\dfrac{Z_{in}(p)}{R_g} + 1}$$

so that

$$Z_{in}(p) \to \infty \text{ or } 0$$

In other words either $Z_{in}(p)$ or $Y_{in}(p)$ has a pole at infinity. Now since all transmission zeros are at infinity we can see that a network of the type shown in Fig. 12.8 is capable of providing the required transfer function since the number of transmission zeros at infinity is equal to the number of reactive elements in the network. Further, the degree of Z_{in} is equal to the number of reactive elements and either Z_{in} or Y_{in} has a pole at infinity depending on whether the first element is a series inductor or a shunt capacitor. Now we can

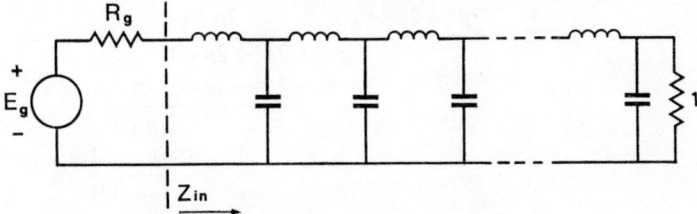

FIG. 12.8. *Ladder network capable of realising any impedance function with transmission zeros at infinity.*

attempt to synthesise Z_{in} or Y_{in} by a continued fraction expansion removing poles at infinity and the degree of the remaining input impedance must be reduced by one after every cycle. To illustrate this let us take the input impedance derived in [*12.34*] and choose the positive sign so that Z_{in} has a pole at infinity. Thus

$$\frac{Z_{in}(p)}{R_g} = \frac{2p^3 + 2p^2 + 2p + 1}{2p^2 + 2p + 1} = \frac{\text{degree 3}}{\text{degree 2}}$$

We now wish to realise this in the form of the ladder network shown in Fig. 12.8. We therefore perform a continued fraction expansion

$$2p^2+2p+1 \overline{\smash{\big)}\ \begin{array}{c} p \\ 2p^3+2p^2+2p+1 \\ \underline{2p^3+2p^2+p} \\ p+1 \end{array}}$$

Thus

$$\frac{Z_{in}}{R_g} = p + \frac{p+1}{2p^2 + 2p + 1}$$

and the remaining impedance is of reduced degree. Notice that in contrast to the case where we performed a continued fraction expansion on a reactance function both numerator and denominator have both even and odd terms and that in consequence one cycle of the continued fraction expansion results in a remainder *two degrees* lower than the dividend, which must be so if the remaining impedance is of reduced degree. The complete expansion for the above example is

$$2p^2+2p+1 \overline{\smash{\big)}\ \begin{array}{c} p \\ 2p^3+2p^2+2p+1 \\ \underline{2p^3+2p^2+p} \\ p+1 \overline{\smash{\big)}\ \begin{array}{c} 2p \\ 2p^2+2p+1 \\ \underline{2p^2+2p} \\ 1 \overline{\smash{\big)}\ \begin{array}{c} p \\ p+1 \\ \underline{p} \\ 1 \overline{\smash{\big)}\ \begin{array}{c} 1 \\ 1 \\ \underline{1} \\ 0 \end{array}} \end{array}} \end{array}} \end{array}}$$

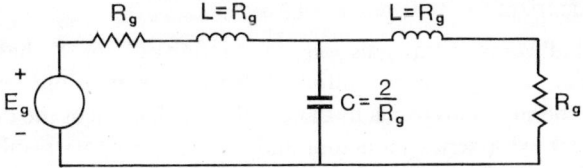

FIG. 12.9. *Realisation of $|S_{12}|^2 = 1/(1+\omega^6)$ with a generator resistance R_g.*

and the network is shown in Fig. 12.9. Note the cancellation of two terms in each cycle. This is a check on the accuracy with which the synthesis has been carried out, and *must* occur in any network with zeros of transmission at infinity.

The synthesis procedure involves only Z_{in}/R_g and it is customary to set $R_g = 1\,\Omega$ since if any other value is subsequently required it can be produced by impedance-scaling. We can therefore write

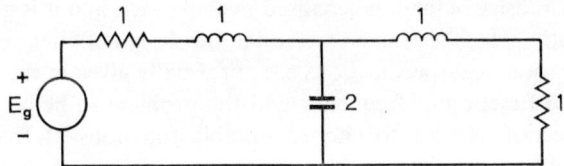

FIG. 12.10. *Realisation of $|S_{12}|^2 = 1/(1+\omega^6)$ with a normalised generator resistance of one ohm.*

Z_{in} in place of Z_{in}/R_g in all relevant formulae. Figure 12.9 is transformed to the circuit of Fig. 12.10 when R_g is set to $1\,\Omega$. If the negative sign were chosen in [*12.34*] then we would have (with $R_g = 1\,\Omega$)

$$Y_{in} = \frac{2p^3 + 2p^2 + 2p + 1}{2p^2 + 2p + 1}$$

and the same continued fraction expansion would apply to Y_{in} yielding the dual circuit of Fig. 12.11.

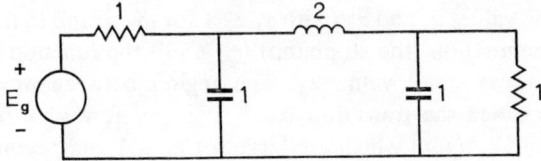

FIG. 12.11. *An alternative realisation of $|S_{12}|^2 = 1/(1+\omega^6)$ with a one ohm generator resistance.*

To summarize:

When all zeros of transmission are at infinity Z_{in} or Y_{in} has a pole at infinity. Carry out a continued fraction expansion removing poles at infinity. Two terms must cancel in each cycle of the division. A network with series inductors and shunt capacitors results. The source has a 1 Ω impedance. If $|S_{12}(0)| = 1$ the load is also 1 Ω, otherwise its value is given by the last (constant) quotient in the continued fraction expansion. Impedance scaling can be used to produce any other source impedance. Either Z_{in} or Y_{in} can be chosen to have the pole at infinity resulting in dual networks.

12.4. Approximation

We have already seen in this chapter how, for example, we may realise prescribed functions for $|Z_{21}(j\omega)|^2$ or $|S_{12}(j\omega)|^2$ using resistively terminated lossless networks. Such functions represent the power in a resistive load, normalised in some way, and it is generally this quantity which is of most interest in the design of filter networks. However, practical specifications are not usually given in the form of a specified function of frequency and the problem to be considered in this section is how to choose suitable functions of frequency, capable of being realised by the required type of network, in order to meet some practical specification. This is known as the approximation problem.

First of all we observe that in consequence of the ideas of frequency transformation and scaling described in Chapter 8, it is only necessary to consider lowpass prototype filters with cutoff frequency at 1 radian per second. The procedures of Chapter 8 then enable us to convert such a prototype into a highpass, bandpass or bandstop filter with an desired bandwidth.

Expressed in terms of the prototype network, the type of specification for the network is generally of the form shown in Fig. 12.12. In the lowpass region (the passband) ($\omega \leq 1$) the function must lie between the values α_1 and α_2 (with $\alpha_1 \leq 1$ for $|S_{12}|$) and in the region of high attentuation (the stopband) ($\omega \geq \omega'$) the function must be less than some small value α_3. The region between $\omega = 1$ and $\omega = \omega'$ is called the transition band. We therefore seek functions $|S_{12}(j\omega)|$ or $|Z_{21}(j\omega)|$ which are large for $\omega \leq 1$ and become small as ω becomes large. The degree of the function determines the

TRANSFER FUNCTION SYNTHESIS

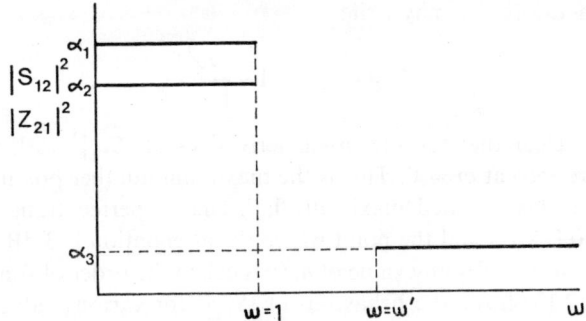

FIG 12.12. *Specification of the attenuation function of a prototype lowpass filter.*

number of circuit elements needed in its realisation, so we would like the performance to improve as the degree is increased. Clearly, we can see from Fig. 12.12 that the severity of the specification is increased as α_2 approaches α_1, or as α_3 is made smaller or as ω' approaches 1 and in general we shall find that any such increase in severity increases the degree of the resulting network.

12.4.1. MAXIMALLY FLAT APPROXIMATIONS

One of the simplest, but none-the-less effective classes of approximating functions is the maximally flat or Butterworth type. In this case we choose

$$|S_{12}(j\omega)|^2 \quad \text{or} \quad |Z_{21}(j\omega)|^2 = \frac{k}{1+\omega^{2n}} \quad [12.35]$$

Henceforth we shall concentrate on $|S_{12}|^2$ and the results for $|Z_{21}|^2$ will be obvious by implication. Now in the case of $|S_{12}|^2$ we shall assume it is desired that all the power from the generator reaches the load at $\omega = 0$ so that $k = 1$ and $\alpha_1 = 1$ in Fig. 12.12. Thus

$$|S_{12}(j\omega)|^2 = \frac{1}{1+\omega^{2n}} \quad [12.36]$$

and

$$|S_{12}(0)| = 1 \quad [12.37]$$

$$|S_{12}(j1)|^2 = \tfrac{1}{2} \quad [12.38]$$

independent of n.

Furthermore, we may write

$$|S_{12}(j\omega)|^2 = 1 - \frac{\omega^{2n}}{1+\omega^{2n}}$$

and it is clear that the first $n-1$ derivatives of $|S_{12}|^2$ with respect to ω^2 are zero at $\omega = 0$. This is the maximum number possible and hence this class is called 'maximally flat'. There is perfect transmission at $\omega = 0$ [*12.37*] and the point where the attenuation is 3 dB always occurs at $\omega = 1$ for any value of n. (n is called the order of function). Figure 12.13 shows the behaviour of $|S_{12}|^2$ for various values of n, from which we see that as n is increased the passband becomes flatter and the attenuation in the stopband at any particular frequency increases. This is therefore the correct type of function for our purposes.

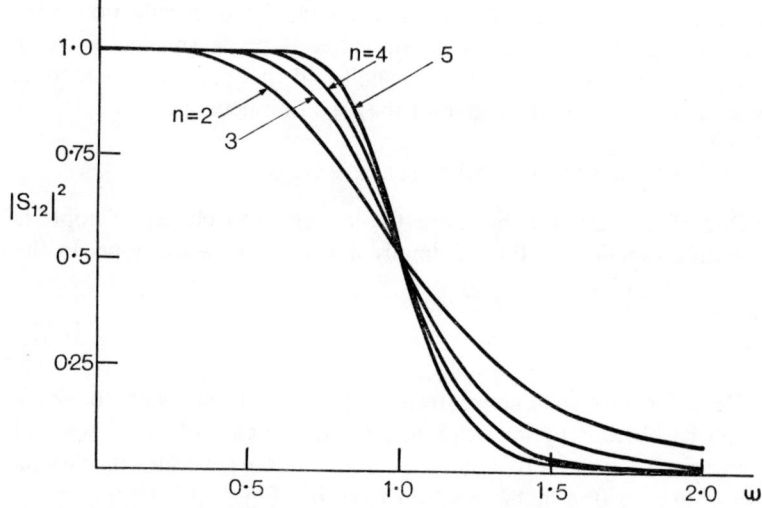

FIG. 12.13. *Maximally flat transducer power gain for $n = 2, 3, 4, 5$.*

Suppose, for example, we specify the 3 dB attenuation at $\omega = 1$ and we require $|S_{12}|^2 \leq \alpha_3$ at $\omega = \omega'$ as in Fig. 12.12. This enables us to determine the minimum value of n since we require

$$\frac{1}{1+(\omega')^{2n}} \leq \alpha_3$$

or
$$(\omega')^{2n} \geq \frac{1}{\alpha_3} - 1 \qquad [12.39]$$

and with $\omega' > 1$ the left hand side increases with increasing n and so any such specification can be met by making n sufficiently large.

We know from Sections 12.2.1 and 12.3 that it is necessary to factorise the denominator of the function whether it represents $|Z_{21}|^2$ or $|S_{12}|^2$. Let us consider, therefore, the realisation of $|S_{12}|^2$, as given by [12.36], in the form of a resistively terminated lossless two-port of the type shown in Fig. 12.4. We assume that the load resistor is 1 ohm and since $|S_{12}(0)| = 1$ the generator resistance must also be 1 ohm.

Following the steps outlined in Section 12.3 we form $|S_{11}(j\omega)|^2$ as

$$|S_{11}(j\omega)|^2 = 1 - |S_{12}(j\omega)|^2$$
$$= \frac{\omega^{2n}}{1+\omega^{2n}} \qquad [12.40]$$

Then
$$S_{11}(p)S_{11}(-p) = \frac{(-p^2)^n}{1+(-p^2)^n} \qquad [12.41]$$

We must now factorise numerator and denominator. The roots of the numerator all occur at $p = 0$ and so require no further consideration. We must therefore find the roots of

$$1+(-p^2)^n = 0$$

or
$$(-p^2)^n = -1 \qquad [12.42]$$

The n roots all have modulus unity so we put

$$-p^2 = \exp(j\theta)$$

and therefore
$$\exp(jn\theta) = -1 \qquad [12.43]$$

or
$$n\theta = (2m-1)\pi \quad m = 1, 2, 3 \ldots \qquad [12.44]$$

Thus
$$-p^2 = \exp\{j(2m-1)\pi/n\} \qquad [12.45]$$

and
$$p = \exp\{j(2m-1+n)\pi/2n\} \quad m = 1, 2, 3 \ldots \quad [12.46]$$
since $-1 = \exp(j\pi)$.

The roots given in [12.46] lie on the unit circle in the complex frequency plane, since their modulus is unity, and they are in turn:

$$\begin{aligned}
p_1 &= \exp\{j(n+1)\pi/2n\} & m &= 1 \\
p_2 &= \exp\{j(n+3)\pi/2n\} & m &= 2 \\
p_3 &= \exp\{j(n+5)\pi/2n\} & m &= 3
\end{aligned}$$

.

.

.

$$\begin{aligned}
p_n &= \exp\{j(3n-1)\pi/2n\} & m &= n \\
p_{n+1} &= \exp\{j(3n+1)\pi/2n\} & m &= n+1
\end{aligned}$$

.

.

.

$$p_{2n} = \exp\{j(5n-1)\pi/2n\} \quad m = 2n$$

any further increase in m merely causes the roots to repeat. Figure 12.14 shows a sketch of such roots. We note that the roots are spaced at equiangular intervals where the angle is π/n and that there are n roots in the left-half plane and n in the right-half plane. The first root is spaced at an angle $\pi/2n$ from the vertical axis. A pair of roots will fall on the real axis at $p = \pm 1$ if for some value of m

$$(2m-1+n)\pi/2n = r\pi$$

or

$$m = \frac{(2r-1)n+1}{2}$$

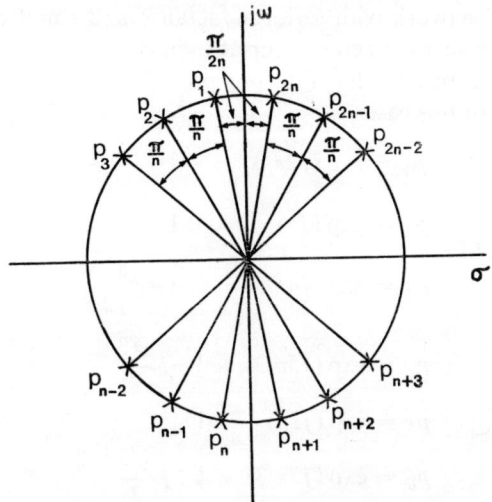

FIG. 12.14. *Pole locations for a maximally flat transducer power gain.*

But since m is an integer this will occur only if n is odd. Thus for odd orders roots will occur at $p = \pm 1$, while if the order is even there will be no real roots. It is also clear that the complex roots occur in conjugate pairs as required to ensure that $S_{12}(p)$ is a real function.

Returning now to the synthesis procedure we can express the denominator of [12.41] as the product of its factors and we choose the factors corresponding to left-half plane roots to form $S_{11}(p)$. Thus

$$S_{11}(p)S_{11}(-p) = \frac{(-p^2)^n}{\prod\limits_{1}^{n}(p-p_i)\prod\limits_{n+1}^{2n}(p-p_j)}$$

where, according to [12.47], the p_i lie in the left-half plane, the p_j lie in the right-half plane.

We then form

$$S_{11}(p) = \frac{\pm p^n}{\prod\limits_{1}^{n}(p-p_i)}$$

$Z_{in}(p)$ can then be formed and synthesised. It is also clear from Section 12.3.1 that this class of functions can be realised in the form of a ladder network with series inductance and shunt capacitance, since all transmission zeros occur at infinity.

It is instructive to illustrate the procedure by an example for, say, $n = 3$. In this case

$$p_1 = \exp(j2\pi/3) = -\tfrac{1}{2} + j\frac{\sqrt{3}}{2}$$

$$p_2 = \exp(j\pi) = -1$$

$$p_3 = \exp(j4\pi/3) = -\tfrac{1}{2} - j\frac{\sqrt{3}}{2}$$

$$p_4 = \exp(j5\pi/3) = \tfrac{1}{2} - j\frac{\sqrt{3}}{2}$$

$$p_5 = \exp(j2\pi) = 1$$

$$p_6 = \exp(j7\pi/3) = \tfrac{1}{2} + j\frac{\sqrt{3}}{2}$$

[*12.47*]

These are shown in Fig. 12.15. Thus we form

$$S_{11}(p) = \frac{\pm p^3}{(p+1)\left(p+\tfrac{1}{2}-j\frac{\sqrt{3}}{2}\right)\left(p+\tfrac{1}{2}+j\frac{\sqrt{3}}{2}\right)}$$

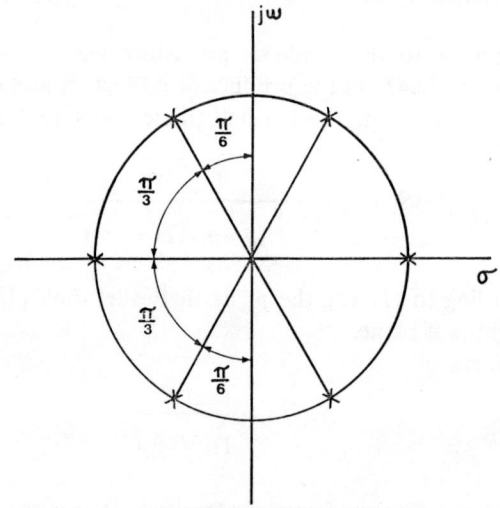

FIG. 12.15. *Pole locations for a maximally flat transducer power gain with $n = 3$.*

TRANSFER FUNCTION SYNTHESIS

by taking the left-half roots in the denominator. Rewriting

$$S_{11}(p) = \frac{\pm p^3}{p^3 + 2p^2 + 2p + 1}$$

and with $R_g = 1$ ohm

$$Z_{in}(p) = \frac{p^3 \pm p^3 + 2p^2 + 2p + 1}{p^3 \mp p^3 + 2p^2 + 2p + 1}$$

from [*12.32*]. We recognise this as the example used in Section 12.31 Taking the positive sign yields the network shown in Fig. 12.10, while choosing the negative sign gives the network of Fig. 12.11.

In the case $n = 4$ we have

$$|S_{12}(j\omega)|^2 = \frac{1}{1+\omega^8}$$

$$|S_{11}(j\omega)|^2 = \frac{\omega^8}{1+\omega^8}$$

$$S_{11}(p)S_{11}(-p) = \frac{p^8}{1+p^8}$$

The roots of the denominator are shown in Fig. 12.16 and are

$$p = \pm \sin(\pi/8) \pm j \cos(\pi/8)$$

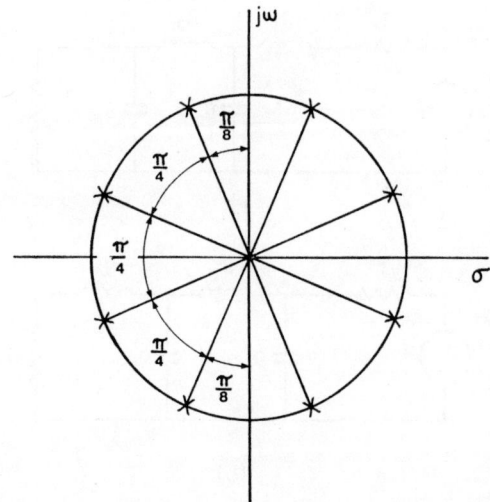

FIG. 12.16. *Pole locations for a maximally flat transducer power gain with $n = 4$.*

and
$$p = \pm \sin(3\pi/8) \pm j\cos(3\pi/8)$$

Selecting left-half plane roots to form $S_{11}(p)$ we get

$$S_{11}(p) = \frac{\pm p^4}{[p^2 + 2p\sin(\pi/8) + 1][p^2 + 2p\sin(3\pi/8) + 1]}$$

$$= \frac{\pm p^4}{p^4 + 2[\sin(\pi/8) + \sin(3\pi/8)]p^3 + 2[1 + \sin(\pi/4)]p^2 + 2[\sin(\pi/8) + \sin(3\pi/8)]p + 1}$$

so that

$$Z_{in}(p) = \frac{p^4 \pm p^4 + 2[\sin(\pi/8) + \sin(3\pi/8)]p^3 + 2[1 + \sin(\pi/4)]p^2 + 2[\sin(\pi/8) + \sin(3\pi/8)]p + 1}{p^4 \mp p^4 + 2[\sin(\pi/8) + \sin(3\pi/8)]p^3 + 2[1 + \sin(\pi/4)]p^2 + 2[\sin(\pi/8) + \sin(3\pi/8)]p + 1}$$

This can then by synthesised by a continued fraction expansion to give the network of Fig. 12.17(a) when the positive sign is taken, and that of Fig. 12.17(b) when the negative sign is used, in forming $S_{11}(p)$.

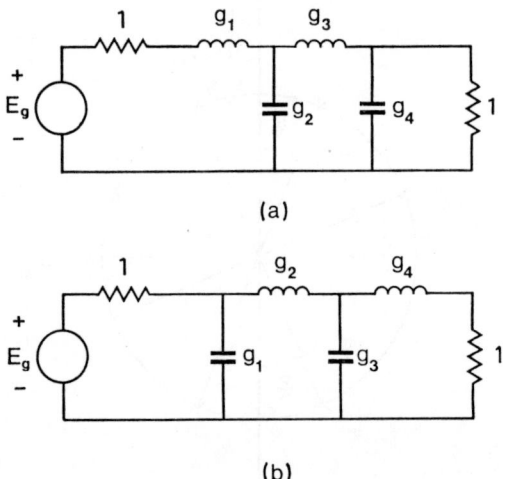

FIG. 12.17. Realisations of 4th order maximally flat transducer power gain; $g_1 = g_4 = 2\sin(\pi/8)$, $g_2 = g_3 = 2\sin(3\pi/8)$.

It has been found possible to derive general formulae for the element values of a ladder network to realise a maximally flat insertion loss, using a network of the type shown in Fig. 12.18(a) or (b). If the network begins with an inductor (capacitor), then it ends with a capacitor (inductor) if n is even, and with an inductor

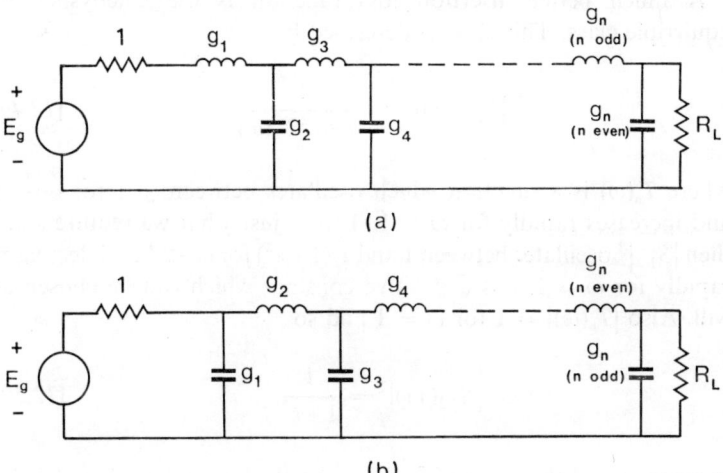

FIG. 12.18. *General realisations of a maximally flat transducer power gain.*

(capacitor) if n is odd. The two forms of network result from choice of the positive or negative sign respectively in the numerator of $S_{11}(p)$. The formulae are

$$g_r = 2 \sin \{(2r-1)\pi/2n\}$$
$$r = 1, 2, 3....n \qquad [12.48]$$

The proof of these formulae is beyond the scope of this book and the reader is well advised to use direct-synthesis techniques in solving any particular problem and to employ the formulae to check his results. R_L is the terminating resistor and in this maximally flat response is always unity.

12.4.2. CHEBYSHEV APPROXIMATIONS

Although the maximally flat type of approximation is widely used it is not generally optimum in meeting specifications of the type given in Fig. 12.12. Basically, this is because the maximally flat type of function is best near the origin (where it meets the specification easily) and not so good near band edge, since we know $|S_{12}|^2$ must be one half at $\omega = 1$. This has the general effect of reducing the rate at which it decreases outside the passband.

A much better insertion loss function is the Chebyshev or equirriple class. This class is described by

$$|S_{12}(j\omega)|^2 = \frac{1}{1+\varepsilon^2 T_n^2(\omega)} \qquad [12.49]$$

where $T_n(\omega)$ is a function which oscillates between ± 1 for $\omega < 1$ and increases rapidly for $\omega > 1$. This is just what we require since then $|S_{12}|^2$ oscillates between 1 and $1/(1+\varepsilon^2)$ for $\omega < 1$ and decreases rapidly for $\omega > 1$. ε is a positive constant which can be chosen at will. Also $|T_n(\omega)| = 1$ for $\omega = 1$ and so

$$|S_{12}(j1)|^2 = \frac{1}{1+\varepsilon^2} \qquad [12.50]$$

independent of the value of n. However, the larger the value of n the smaller the value of $|S_{12}|$ for any ω greater than unity so that the stopband performance improves with increasing n. The function $T_n(\omega)$ has the property that $T_n(0) = 0$ for n odd and $T_n(0) = 1$ for n even. The general behaviour of a Chebyshev response is shown in Fig. 12.19.

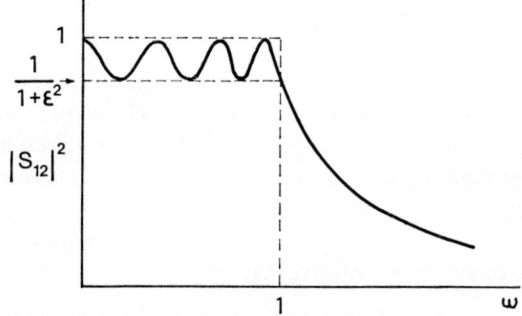

FIG. 12.19. *General form of the transducer power gain* $1/\{1+\varepsilon^2 T_n^2(\omega)\}$.

TRANSFER FUNCTION SYNTHESIS

In terms of the type of specification given in Fig. 12.12 (with $\alpha_1 = 1$) we know that at $\omega = 1$, $|T_n(\omega)| = 1$, so that

$$\frac{1}{1+\varepsilon^2} = \alpha_2 \qquad [12.51]$$

and

$$\varepsilon^2 = \frac{1}{\alpha_2} - 1 \qquad [12.52]$$

independent of n. Thus the passband specification can be met by *any* value of n. For the stopband we have

$$\frac{1}{1+[(1/\alpha_2)-1]T_n^2(\omega')} = \alpha_3 \qquad [12.53]$$

and if we know the exact form of $T_n(\omega)$ for various n we can select the appropriate one to satisfy this equation.

We must now consider the exact form of the function $T_n(\omega)$. This is defined as

$$T_n(\omega) = \cosh(n \cosh^{-1}\omega)$$
$$= \cos(n \cos^{-1}\omega) \qquad [12.54]$$

It is easy to see that these two definitions are the same if we put

$$x = \cosh^{-1}\omega$$
$$y = \cos^{-1}\omega \qquad [12.55]$$

so that

$$\omega = \cosh x = \cos y \qquad [12.56]$$

Then

$$T_n(\omega) = \cosh nx = \frac{\exp(nx) + \exp(-nx)}{2} \qquad [12.57]$$

$$= \cos ny = \frac{\exp(jny) + \exp(-jny)}{2}$$

Thus from [12.57] the definitions are the same if $x = jy$. But from [12.56]

$$\cosh x = \cos y$$

and putting $x = jy$ we get

$$\cosh x = \cosh(jy) = \cos y$$

so the two forms given in [12.54] are identical and may be interchanged in any convenient way.

To see that [12.54] has the properties already ascribed to $T_n(\omega)$ consider the second form. For $-1 \leq \omega \leq 1$, $\cos^{-1}\omega$ is a real angle and as ω varies from -1 to 0 to 1, $\cos^{-1}\omega$ varies from π to $\pi/2$ to 0. Thus in the same interval $n\cos^{-1}\omega$ varies from $n\pi$ to 0. Thus $T_n(\omega)$ which is the cosine of this argument varies between ± 1 and passes through $n/2$ cycles. This behaviour is illustrated for $n = 3$ and $n = 4$ in Fig. 12.20. Note that when $\omega = 1$, $\cos^{-1}\omega = 0$ and so $T_n(\omega) = 1$ independent of n. Also, when $\omega = 0$, $T_n(\omega) = \cos(n\pi/2)$ which is zero for n odd and is ± 1 for n even.

For $|\omega| > 1$ the first form of [12.54] is the more convenient. In this case $\cosh^{-1}\omega$ is a real number which increases with increasing ω and is greater than unity. Thus $\cosh(n\cosh^{-1}\omega)$ increases with increasing ω for a given value of n and increases with increasing n for a given value of ω (> 1).

To summarise: we see that $T_n(\omega)$ oscillates between ± 1 for $|\omega| \leq 1$, is always unity when $\omega = 1$, increases with increasing ω or n for $|\omega| > 1$ and is zero at $\omega = 0$ for n odd and ± 1 at $\omega = 0$ for n even. Also $T_n(\omega)$ passes through $n/2$ cycles between $\omega = 1$ and $\omega = -1$. Figure 12.21 shows the behaviour of $T_3(\omega)$ over an extended range.

One further important point remains. If [12.49] is to be realisable as a transducer power gain then $T_n(\omega)$ must be a polynomial function of ω and we must demonstrate that this is so by reference to the definition given in [12.54]. To do so we observe that with $x = \cosh^{-1}\omega$

$$T_n(\omega) = \cosh nx$$

$$= \sum_0^k (-1)^r \frac{n}{n-r} \binom{n-r}{r} 2^{n-2r-1}(\cosh x)^{n-2r}$$

where $k = n/2$ for n even and $k = (n-1)/2$ for n odd. But $\cosh x = \omega$ so

$$T_n(\omega) = \sum_0^k (-1)^r \frac{n}{n-r} \binom{n-r}{r} 2^{n-2r-1}\omega^{n-2r} \qquad [12.58]$$

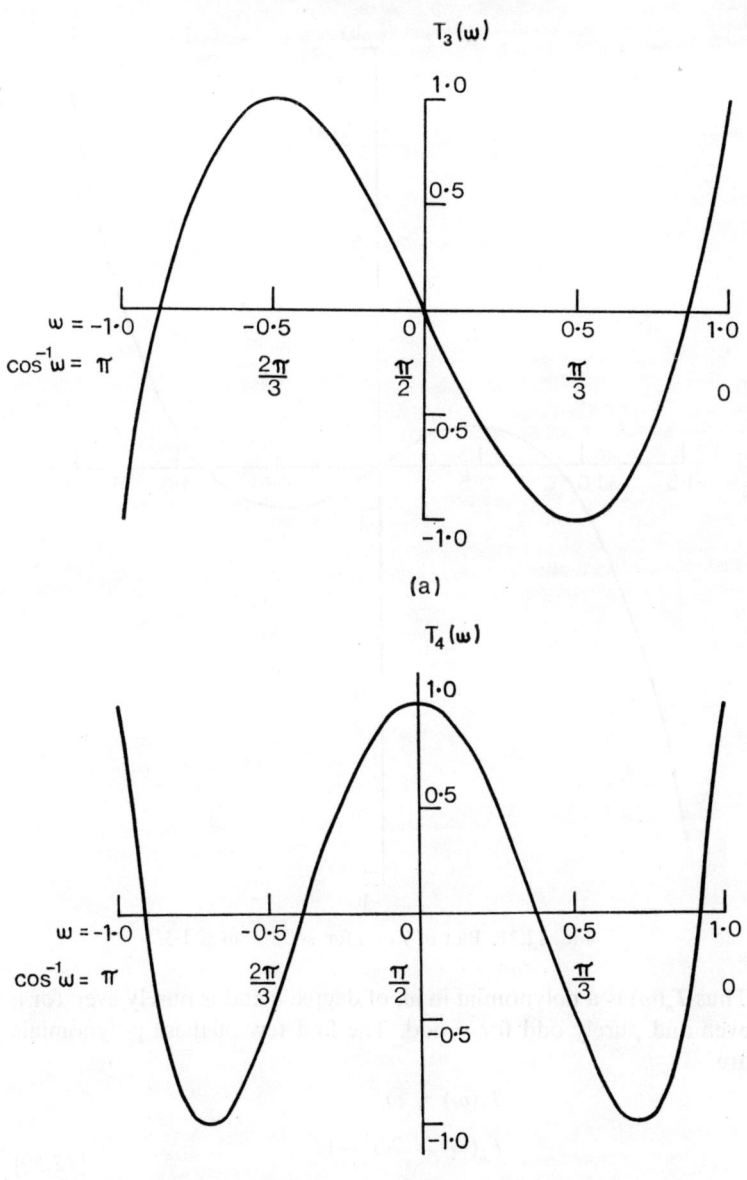

FIG. 12.20. *Plot of $T_n(\omega)$ for $-1 \leqq \omega \leqq 1$.* (a) $n = 3$; (b) $n = 4$.

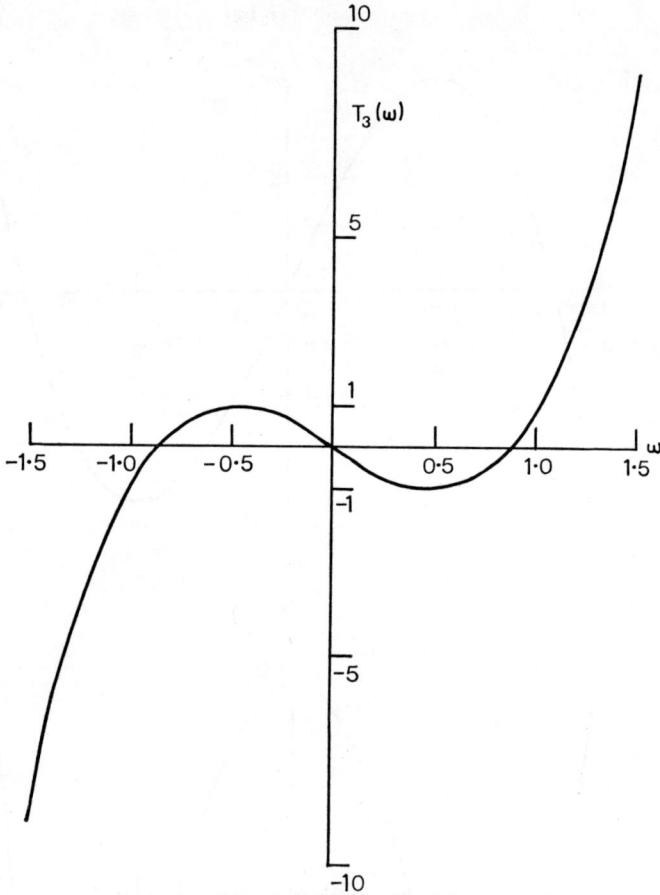

Fig. 12.21. Plot of $T_3(\omega)$ for $-1{\cdot}5 \leqq \omega \leqq 1{\cdot}5$.

Thus $T_n(\omega)$ is a polynomial in ω, of degree n and is purely even for n even and purely odd for n odd. The first few of these polynomials are

$$T_1(\omega) = \omega$$
$$T_2(\omega) = 2\omega^2 - 1$$
$$T_3(\omega) = 4\omega^3 - 3\omega$$
$$T_4(\omega) = 8\omega^4 - 8\omega^2 + 1$$

[*12.59*]

TRANSFER FUNCTION SYNTHESIS

A simple recursion formula enables higher order functions to be easily calculated. This is

$$T_{n+1}(\omega) - 2\omega T_n(\omega) + T_{n-1}(\omega) = 0 \qquad [12.60]$$

Also the square of a Chebyshev polynomial is equivalent (apart from an additive constant) to one of double order.

$$T_n^2(\omega) = \tfrac{1}{2}\{T_{2n}(\omega) + 1\} \qquad [12.61]$$

Thus using [12.53] we can choose an appropriate polynomial, $T_n(\omega)$, to achieve any desired stopband performance. It can also be shown that the transducer power gain of [12.49] produces the fastest rate of cutoff in the stopband compared to any other function where $|S_{12}|^2$ has a constant numerator and meets the same passband specification.

If we now consider the realisation of [12.49] we first form

$$|S_{11}(j\omega)|^2 = \frac{\varepsilon^2 T_n^2(\omega)}{1 + \varepsilon^2 T_n^2(\omega)} \qquad [12.62]$$

and

$$S_{11}(p)S_{11}(-p) = \frac{\varepsilon^2 T_n^2(p/j)}{1 + \varepsilon^2 T_n^2(p/j)} \qquad [12.63]$$

putting $j\omega = p$.

We must now factorise numerator and denominator, in order to form $S_{11}(p)$. Considering first the denominator we require the roots of

$$1 + \varepsilon^2 T_n^2(p/j) = 0$$

or

$$T_n(p/j) = \pm j/\varepsilon$$

This gives

$$\cosh\{n \cosh^{-1}(p/j)\} = \pm j/\varepsilon$$

or

$$n \cosh^{-1}(p/j) = \cosh^{-1}(\pm j/\varepsilon)$$

or

$$p/j = \cosh[(1/n)\cosh^{-1} \pm j/\varepsilon] \qquad [12.64]$$

Now putting

$$\alpha + j\beta = \cosh^{-1}[\pm (j/\varepsilon)]$$

$$\cosh(\alpha + j\beta) = \pm (j/\varepsilon)$$

$$\cosh \alpha \cos \beta + j \sinh \alpha \sin \beta = \pm (j/\varepsilon)$$

Equating real and imaginary parts

$$\cosh \alpha \cos \beta = 0$$

$$\sinh \alpha \sin \beta = \pm (1/\varepsilon)$$

Thus

$$\beta = (2m-1)\pi/2 \qquad m = 1, 2, 3, \ldots$$

$$\sinh \alpha = \pm (1/\varepsilon)$$

Thus

$$\alpha + j\beta = \pm \sinh^{-1}(1/\varepsilon) + j(2m-1)\pi/2$$

and

$$\begin{aligned}
p &= j \cosh\left[\pm \frac{1}{n} \sinh^{-1}\left(\frac{1}{\varepsilon}\right) + j \frac{(2m-1)\pi}{2n}\right] \\
&= j\left[\cosh\left\{\frac{1}{n} \sinh^{-1}\left(\frac{1}{\varepsilon}\right)\right\} \cos\left(\frac{(2m-1)\pi}{2n}\right) \right. \\
&\qquad \left. \pm j \sinh\left\{\frac{1}{n} \sinh^{-1}\left(\frac{1}{\varepsilon}\right)\right\} \sin\left(\frac{(2m-1)\pi}{2n}\right)\right]
\end{aligned}$$

or

$$\begin{aligned}
p &= \sin\left(\frac{(2m-1)\pi}{2n}\right) \sinh\left(\frac{1}{n} \sinh^{-1} \frac{1}{\varepsilon}\right) \\
&\quad + j \cos\left(\frac{2m-1}{2n} \pi\right) \cosh\left(\frac{1}{n} \sinh^{-1} \frac{1}{\varepsilon}\right) \qquad m = 1, 2, 3, \ldots
\end{aligned}$$

[12.65]

where we have ignored the $\pm$ signs since a sufficient range of m will generate all possible roots in any case.

Let us now consider the location of the roots given in [*12.65*]. They may be written as

$$p_i = \sigma_i + j\omega_i; \quad i = 1, 2, 3, \ldots \qquad [12.66]$$

and

$$\frac{\sigma_i^2}{\sinh^2[(1/n)\sinh^{-1}(1/\varepsilon)]} + \frac{\omega_i^2}{\cosh^2[(1/n)\sinh^{-1}(1/\varepsilon)]} = 1 \qquad [12.67]$$

so that the roots lie on an ellipse in the complex frequency plane. Figure 12.22 shows the location of the roots where the angles are equal to those of the corresponding roots in the maximally flat case as is evident from [*12.65*]. The semi-minor and semi-major axes of the

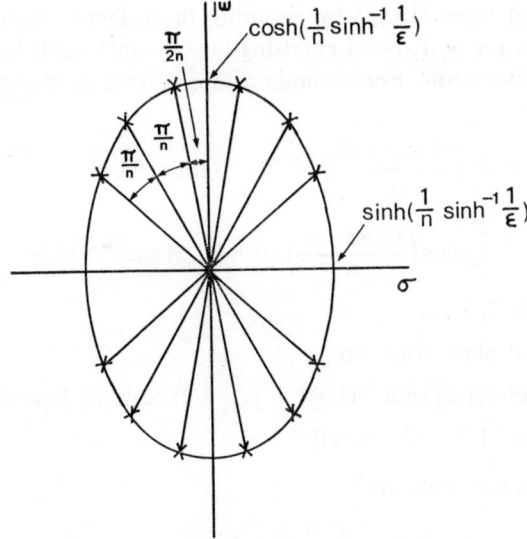

FIG. 12.22. *Pole locations for a Chebyshev transducer power gain.*

ellipse are $\sinh[(1/n)\sinh^{-1}(1/\varepsilon)]$ and $\cosh[(1/n)\sinh^{-1}(1/\varepsilon)]$ respectively. There are, of course, n left-half plane and n right-half plane roots, and when n is odd there are a pair of roots at $p = \pm \sinh[(1/n)\sinh^{-1}(1/\varepsilon)]$. All other roots occur in conjugate pairs.

We must now find the roots of the numerator of [*12.63*] given by

$$T_n^2(p/j) = 0$$

or

$$\cos(n \cos^{-1} p/j) = 0$$

so that

$$n \cos^{-1}(p/j) = (2r-1)(\pi/2)$$

$$\frac{p}{j} = \cos\frac{(2r-1)\pi}{2n} \quad r = 1, 2, 3\ldots$$

and

$$p = j\cos\frac{(2r-1)\pi}{2n} \quad r = 1, 2, 3\ldots$$

[*12.68*]

Thus the roots of the numerator are purely imaginary. We can now form $S_{11}(p)$ from [*12.63*] by choosing those factors with left-half plane roots for $S_{11}(p)$ and choosing any suitable (real) selection of the numerator roots. For example with $n = 3$ we have denominator roots

$$p = \sin\left(\frac{(2m-1)\pi}{6}\right) \sinh\,[(1/3)\sinh^{-1}(1/\varepsilon)]$$

$$+ j\cos\left(\frac{(2m-1)\pi}{6}\right) \cosh\,[(1/3)\sinh^{-1}(1/\varepsilon)]$$

with $m = 1, 2, 3 \ldots$

The left-half plane roots are

$p = -\tfrac{1}{2}\sinh\,[(1/3)\sinh^{-1}(1/\varepsilon)] \pm j\,(\sqrt{3}/2)\cosh\,[(1/3)\sinh^{-1}(1/\varepsilon)]$
$p = -\sinh\,[(1/3)\sinh^{-1}(1/\varepsilon)]$ [*12.69*]

The numerator roots are

$$p = \pm j(\sqrt{3}/2)$$
$$p = 0 \qquad [12.70]$$

Thus

$S_{11}(p) = [\pm p(p^2 + \tfrac{3}{4})]/\{p + \sinh\,[(1/3)\sinh^{-1}(1/\varepsilon)]\}$

$\times \{p^2 + p\sinh\,[(1/3)\sinh^{-1}(1/\varepsilon)] + \tfrac{1}{4}\sinh^2[(1/3)\sinh^{-1}(1/\varepsilon)]$

$+ \tfrac{3}{4}\cosh^2[(1/3)\sinh^{-1}(1/\varepsilon)]\}$ [*12.71*]

TRANSFER FUNCTION SYNTHESIS

where we have ensured that the coefficients of the highest powers in p are equal to unity in numerator and denominator. (Note that in [12.63] they are equal but not unity and can be made unity by dividing numerator and denominator by a suitable number.) We now form $Z_{in}(p)$ as before and find two networks of the type shown in Figs. 12.10 and 12.11 (with different element values, of course) corresponding to the choice of the positive and negative signs in [12.71].

Again, in general, ladder networks of the type shown in Fig. 12.18 result from Chebyshev responses and again general formulae for the element values in Fig. 12.18 have been found. The proof is beyond the scope of this book. The formulae are

$$g_1 = \frac{2 \sin(\pi/2n)}{\sinh[(1/n)\sinh^{-1}(1/\varepsilon)]}$$

$$g_r g_{r+1} = \frac{4 \sin\{(2r-1)\pi/2n\} \sin\{(2r+1)\pi/2n\}}{\sinh^2[(1/n)\sinh^{-1}(1/\varepsilon)] + \sin^2(r\pi/n)} \quad [12.72]$$

where $r = 1, 2, 3 \ldots n-1$.

In the case where n is odd we know that $S_{12}(0) = 1$ and so the terminating resistor is one ohm. However, when n is even $|S_{12}(0)|^2 = 1/(1+\varepsilon^2)$ and so the terminating resistor is not unity. At $\omega = 0$

$$|S_{12}|^2 = \frac{4R_L}{(R_L+1)^2} \quad [12.73]$$

and so R_L may be determined in terms of ε for n even.

This gives

$$R_L = [\varepsilon + \sqrt{(1+\varepsilon^2)}]^2 \quad (a)$$

or

$$R_L = \left(\frac{1}{[\varepsilon + \sqrt{(1+\varepsilon^2)}]}\right)^2 \quad (b)$$

[12.74]

This choice depends on whether $S_{11}(0)$ is chosen to be positive or negative. For $S_{11}(0) > 0$, [12.74a] is chosen ($R_L > 1$), while [12.74b] gives $S_{11}(0) < 0$ ($R_L < 1$).

12.4.3. BANDPASS APPROXIMATIONS

In the two previous sub-sections, we have seen how maximally flat and Chebyshev approximations for lowpass prototype filters with cutoff frequency $\omega = 1$ may be derived. If a lowpass filter with any other cutoff frequency is specified, one merely frequency-scales the lowpass prototype according to the methods given in Chapter 8. In the case of a bandpass characteristic the specification is usually given in the form shown in Fig. 12.23. In the case of a maximally flat approximation, we assume for convenience that $\alpha_1 = 0.5$, while in the case of a Chebyshev approximation $\alpha_1 = 1/(1+\varepsilon^2)$. Now we

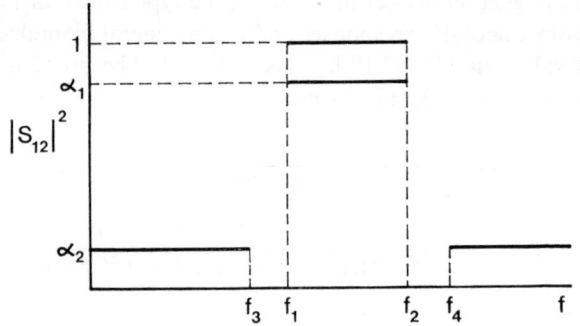

FIG. 12.23. *General form of a bandpass transducer power gain specification.*

know that the design procedure, as outlined in Section 8.2.4, ensures that the centre frequency of the characteristic is the geometric mean of any two frequencies with the same level of attentuation. Thus $f_0 = (f_1 f_2)^{\frac{1}{2}}$. Now we have, in general, no assurance that f_3 and f_4 are related in such a way that

$$f_3 f_4 = f_1 f_2 = f_0^2$$

so we must choose a characteristic which reaches the level α_2 at the more severe of the specified frequencies f_3 and f_4, and we are then assured that the overall specification is satisfied. This is clarified by an example. Suppose $f_1 = 1$ MHz, $f_2 = 1.44$ MHz, $f_3 = 0.7$ MHz, $f_4 = 2.2$ MHz, then

$$f_0 = \sqrt{(f_1 f_2)} = 1.2 \text{ MHz}$$

Now suppose we choose a characteristic which has the value α_2 at f_3, then the value α_2 will also be attained at some frequency f_4' where

$$f_3 f_4' = f_0^2$$

or
$$f_4' = \frac{1.44}{0.7} = 2.057 \text{ MHz}$$

This is lower than the specified f_4 so that the specification will be met in this respect. On the other hand, if we had chosen a characteristic which only just achieved the level α_2 at f_4 it would have failed to do so at f_3.

Now, given the bandpass specification, we must choose the appropriate lowpass prototype. If the characteristic is maximally flat then only the degree needs to be chosen by reference to the stopband specification, while if the characteristic is of the Chebyshev type, ε is determined from the value of α_1 and the order is chosen by reference to the more severe of the stopband specifications.

If, in general terms, the lowpass prototype has a transducer power gain

$$|S_{12}|^2 = F_n(\omega)$$

having the required passband behaviour and cutoff frequency $\omega = 1$, then from Section 8.2.4 a similar bandpass characteristic of bandwidth unity is

$$|S_{12}|^2 = F_n\left(\omega - \frac{\omega_0^2}{\omega}\right)$$

while a similar bandpass characteristic with bandwidth $\omega_2 - \omega_1$ is

$$|S_{12}|^2 = F_n\left\{\frac{\omega_0}{\omega_2 - \omega_1}\left(\frac{\omega}{\omega_0} - \frac{\omega_0}{\omega}\right)\right\}$$

where $\omega_0^2 = \omega_1 \omega_2$.

Now if, as before, we assume that f_3 gives the more severe stopband specification, we require

$$F_n\left\{\frac{\omega_0}{\omega_2 - \omega_1}\left(\frac{\omega_3}{\omega_0} - \frac{\omega_0}{\omega_3}\right)\right\} \leqq \alpha_2$$

and the order of $|S_{12}|$ must be chosen accordingly. One then knows the lowpass prototype function which is realised as explained previously. This is then frequency-scaled, resonated and impedance-scaled to give the required bandpass characteristic.

As an example, consider the following specification. f_1, f_2, f_3, f_4 have the values previously given. A maximally flat characteristic (with $\alpha_1 = 0.5$) is required with $\alpha_2 = 0.0012$, and 50-ohm terminations.

The bandpass characteristic is therefore

$$|S_{12}|^2 = \frac{1}{1+\left(\dfrac{\omega_0}{\omega_2-\omega_1}\right)^{2n}\left(\dfrac{\omega}{\omega_0}-\dfrac{\omega_0}{\omega}\right)^{2n}}$$

and, since the specification at 0·7 MHz is the more severe stopband requirement, we must have

$$\frac{1}{1+\left(\dfrac{1.2}{0.44}\right)^{2n}\left(\dfrac{0.7}{1.2}-\dfrac{1.2}{0.7}\right)^{2n}} \leqq 0.0012$$

i.e.
$$1+(3\cdot1)^{2n} \geqq 833$$

or
$$(3\cdot1)^{2n} \geqq 832$$

It is clear that $n = 1$ or $n = 2$ is too small but $n = 3$ gives

$$(3\cdot1)^6 = 888 > 832$$

Thus we require a third degree prototype

$$|S_{12}|^2 = \frac{1}{1+\omega^6}$$

This, as we know, leads to the circuit of Fig. 12.10 or Fig. 12.11. Assume that we use Fig. 12.11. We then frequency-scale to obtain the correct bandwidth of 0·44 MHz so that each capacitor becomes $1/(0.44 . 2\pi . 10^6)$ and the inductor becomes $2/(0.44 . 2\pi . 10^6)$. We can then impedance-scale to obtain 50-ohm terminations so that each capacitor becomes $1/(0.44 . 100\pi . 10^6)$ and the inductor becomes $100/(0.44 . 2\pi . 10^6)$. Finally, we resonate each element (the capacitors by means of parallel inductors, and the inductor by a series capacitor) to the required centre frequency of 1·2 MHz.

12.4.4. DELAY APPROXIMATIONS

The type of specification we have considered so far has been concerned entirely with the insertion loss of the filter. In many appli-

TRANSFER FUNCTION SYNTHESIS

cations the phase response, often expressed in terms of group delay, is also of importance. This is generally a rather more difficult topic and although a maximally flat approximation to a constant group delay has been derived, no corresponding analytic equiripple result has been achieved.

As explained in Section 8.4.3, if we write

$$S_{12}(j\omega) = \alpha(\omega) + j\beta(\omega)$$

then the associated phase angle is

$$\psi(\omega) = \tan^{-1} \frac{\beta(\omega)}{\alpha(\omega)}$$

while the group delay is

$$T_g(\omega) = \frac{d}{d\omega}(\psi)$$

$$= \frac{d}{d\omega} \tan^{-1} \frac{\beta(\omega)}{\alpha(\omega)}$$

$$= \frac{\alpha\beta' - \beta\alpha'}{\alpha^2 + \beta^2} \qquad [12.75]$$

where prime denotes differentiation with respect to ω. In the ideal case we would require the group delay to be constant within the passband and so we seek approximations to this behaviour which are maximally flat at $\omega = 0$ in the lowpass prototype case.

If we restrict the discussion to functions of the type

$$S_{12}(p) = \frac{K}{D_N(p)} \qquad [12.76]$$

$$= \frac{K}{m(p) + n(p)} \qquad [12.77]$$

where m is even, n is odd and $m+n$ is Hurwitz, then we can write

$$T_g(p) = \frac{mn' - nm'}{m^2 - n^2} \qquad [12.78]$$

where prime denotes differentiation with respect to p, such that putting $p = j\omega$ gives the actual group delay.

The problem is now to choose a Hurwitz polynomial $m+n$ such that [12.78] is a maximally flat approximation to a constant at $\omega = 0$. To achieve this, consider the (unrealisable) function

$$S_{12}(p) = \frac{K}{\cosh(ap) + \sinh(ap)} \qquad [12.79]$$

so that

$$m = \cosh(ap)$$
$$n = \sinh(ap)$$

and [12.78] gives

$$T_g(p) \equiv a \qquad [12.80]$$

Thus in this case the delay is a constant independent of frequency, so we should choose m and n to approximate this function in some way. A maximally flat result is obtained if we consider $n(p)/m(p)$ which in [12.79] is $\tanh(ap)$. $n(p)/m(p)$ is a function of degree N so we must set this equal to some approximation to $\tanh(ap)$ of degree N. We must, in fact, choose a continued function expansion of $\tanh(ap)$. This is given as

$$\tanh(ap) = \cfrac{1}{\cfrac{1}{ap} + \cfrac{1}{\cfrac{3}{ap} + \cfrac{1}{\cfrac{5}{ap} + \cdots \cfrac{1}{\cfrac{2N-1}{ap} + \cdots}}}} \qquad [12.81]$$

and is found by expanding $\sinh(ap)$ and $\cosh(ap)$ as infinite power series and then performing a continued function expansion of their quotient. To form n/m we truncate this expansion after N quotients,

TRANSFER FUNCTION SYNTHESIS

remultiply and set the result equal to n/m. The numerator being odd is therefore equal to n while the denominator being even is equal to m. For example, if we require the third degree function, then

$$\tanh(ap) = \cfrac{1}{\cfrac{1}{ap}+\cfrac{1}{\cfrac{3}{ap}+\cfrac{1}{\cfrac{5}{ap}}}}$$

[12.82]

and

$$\frac{n}{m} = \frac{15ap + a^3p^3}{15 + 6a^2p^2}$$

so

$$m = 15 + 6a^2p^2$$
$$n = 15ap + a^3p^3$$

and the delay at $p = 0$ is always a. So

$$S_{12}(p) = \frac{K}{15 + 15ap + 6a^2p^2 + a^3p^3} \qquad [12.83]$$

and we put $K = 15$ to make $S_{12}(0) = 1$.

Since, in the continued fraction expansion of $\tanh(ap)$ the quotients are all positive, the continued fraction expansion of n/m also results in quotients which are all positive and we have seen earlier that this means n/m is a reactance function and hence that $m+n$ is a Hurwitz polynomial. Thus $S_{12}(p)$ is always realisable for some value of K. In fact, if we choose K so that $S_{12}(0) = 1$, then $|S_{12}| \leq 1$ for all ω as required for realisability.

Although $S_{12}(p)$ may be formed directly from [12.81], the remultiplication required is rather tedious and formulae have been derived to generate $m+n$. A recurrence formula is

$$D_N(ap) = (2N-1)D_{N-1}(ap) + a^2p^2 D_{N-2}(ap) \qquad [12.84]$$

with

$$D_1 = 1 + ap, \quad D_2 = 3 + 3ap + a^2p^2$$

These polynomials are called *Bessel Polynomials*, and it has been shown, in general, that the resulting delay is indeed maximally flat.

In the case $N = 3$ considered above, this can readily be shown. If we calculate $T_g(p)$ from [12.78], we obtain

$$T_g(p) = a \frac{225 - 45a^2p^2 + 6a^4p^4}{225 - 45a^2p^2 + 6a^4p^4 - a^6p^6}$$

$$= 1 + \frac{a^6p^6}{225 - 45a^2p^2 + 6a^4p^4 - a^6p^6}$$

which clearly has two derivatives with respect to p^2 equal to zero at $p = 0$ as required. The proof that these functions yield maximally flat delay in all cases is beyond the scope of this book.

It should be mentioned that the attenuation characteristic resulting from this class of functions is rather poor in the sense of its ability to meet specifications of the type given in Fig. 12.12. Also, if the previous lowpass to bandpass transformation is used then the group delay characteristic is distorted as compared to that of the lowpass prototype. However, in the lowpass case, if frequency scaling is applied the delay response is merely scaled (since it is a function only of ap) and the d.c. delay is also scaled accordingly.

12.4.5. MORE GENERAL APPROXIMATIONS

In the previous sections we have seen how a function of the type $K/D(p)$ may be used to approximate amplitude or delay specifications. The general subject of approximation is, however, much broader than this and it is beyond the scope of this book to present a thorough treatment of the subject. However, some general observations are possible, based on what we have already seen. Firstly, it is possible to meet any arbitrary amplitude specification using either a maximally flat or Chebyshev characteristic of appropriate degree. Furthermore, the type of network required is very simple, being a ladder with series inductance and shunt capacitance. One might therefore wonder whether there is a need for anything further. In practice, there are several reasons for investigating other possibilities. Since it is known that the Chebyshev response provides the best possible characteristic when the numerator of S_{12} is a constant, any possible improvement will require a function S_{12} with a polynomial numerator. As we have seen previously, the zeros of S_{12} are called the zeros of transmission and when the numerator is a constant the zeros of transmission occur only at $p = \infty$. With a polynomial numerator we have finite zeros of transmission and the function cannot be realised by a simple ladder network. However, the hope is that in spite of this the resulting network required to meet a given specification may yet be simpler. An example of this class of filters is the so-called elliptic function type. In this case, S_{12} is of the form

$$|S_{12}|^2 = \frac{\pi(p^2+\omega_i^2)^2}{D_n(p)D_n(-p)}$$

so that the finite zeros of transmission occur at real frequencies, ω_i. These, and the denominator, are chosen to give a response of the type shown in Fig. 12.24. This provides equiripple behaviour in both

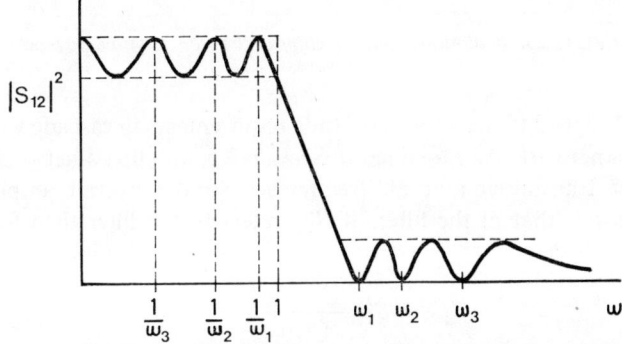

FIG. 12.24. *General form of an elliptic function transducer power gain.*

pass and stop bands, and the degree required for a given specification is less than the corresponding Chebyshev case. These functions may usually be realised by networks of the type shown in Fig. 12.25. The transmission zeros are produced by the resonances of the series or parallel tuned circuits which are appropriately tuned. The detailed form of $|S_{12}|^2$ is given in terms of Jacobian Elliptic functions, and the synthesis procedures are rather cumbersome.

In another common class of filters both the amplitude and delay characteristics are specified. In order to exercise control of both characteristics, finite zeros of transmission must be used and the choice of approximating functions in this case represents one of the most difficult problems in the area of approximation theory. Another approach to this problem is to attempt a realisation in the form of a

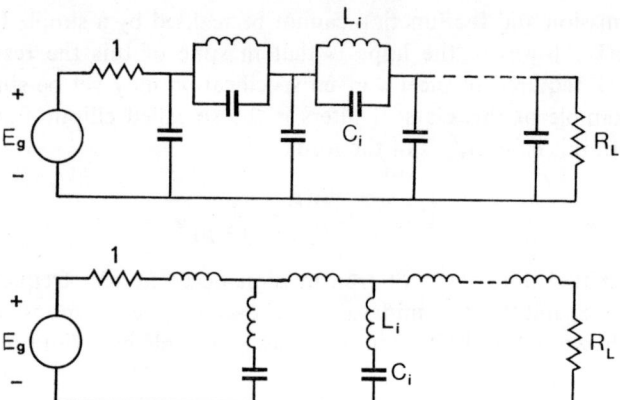

Fig. 12.25. *Realisations of an elliptic function transducer power gain.*

filter designed to meet the amplitude requirements in cascade with an allpass network. As explained in Section 8.5, an allpass network has perfect transmission at all frequencies so the overall amplitude response is that of the filter. If S'_{12} refers to the filter then for the combination

$$S_{12} = S'_{12} \frac{Q(-p)}{Q(p)}$$

and

$$S_{12}(p)S_{12}(-p) = S'_{12}(p)S'_{12}(-p) \frac{Q(-p)Q(p)}{Q(p)Q(-p)}$$

$$= S'_{12}(p)S'_{12}(-p)$$

Thus the amplitude response is determined by the filter alone. Now if $S'_{12}(j\omega) = \alpha(\omega) + j\beta(\omega)$, and if $Q(j\omega) = m(\omega) + jn(\omega)$, then

$$S_{12}(j\omega) = \{\alpha(\omega) + j\beta(\omega)\} \frac{m(\omega) - jn(\omega)}{m(\omega) + jn(\omega)}$$

and the overall phase angle is

$$\psi(\omega) = \tan^{-1}\left(\frac{\beta}{\alpha}\right) + \tan^{-1}\left(\frac{-n}{m}\right) - \tan^{-1}\left(\frac{n}{m}\right)$$

$$= \tan^{-1}\left(\frac{\beta}{\alpha}\right) - 2\tan^{-1}\left(\frac{n}{m}\right)$$

TRANSFER FUNCTION SYNTHESIS   521

$Q(p)$ must then be chosen so that $d\psi/d\omega$ meets the overall delay specification. The delay of the allpass network is the sum of the delays of C and D-sections as explained in Section 8.5, and there is, as yet, no analytic means of choosing these. Numerical procedures must therefore be used.

We have thus seen that the approximation problem has many aspects and that in several important areas elegant optimum analytic results have been obtained. In other cases no such results are yet known and numerical techniques must be used. Active research is in progress on many aspects of the approximation problem.

12.5. Elements of Cascade Synthesis

We have seen in Chapter 9 how the Darlington synthesis procedure for a driving-point impedance leads to the specification of a lossless two-port which is to be terminated at port 2 by a one-ohm resistor in order to realise the prescribed impedance at port 1. We have also seen, in Section 10.2, how the lossless two-port may be realised by expanding its z- or y-parameters in partial fractions; realising each term of the matrix expansion by a simple network, and then connecting the resulting networks in series or parallel. This method is perfectly general but suffers from the practical disadvantages that interconnecting the networks requires many ideal transformers in a high degree network, and the effect of any particular network element on the overall response is not easily isolated. The method of cascade synthesis attempts to overcome these difficulties.

The basis of the method is contained in Section 12.1. There we have seen how a driving-point impedance may be classified in terms of its zeros of transmission. [*12.3*] gives

$$Z(s)+Z(-s) = \frac{-2z_{12}^2(s)}{1-z_{22}^2(s)} \qquad [12.85]$$

for a driving-point impedance $Z(s)$, realised as a lossless network terminated in a one-ohm resistor, as shown in Fig. 12.26. Thus the zeros of $Z(s)+Z(-s)$ are the zeros of $z_{12}^2(s)$ since in the Darlington procedure the lossless network is compact. [*12.4*] shows that at such values of s, $Z_{21}(s)Z_{21}(-s)$ is zero and there is no transmission through the lossless network. Suppose we were able to realise the lossless network in the form of a cascade of simpler lossless networks

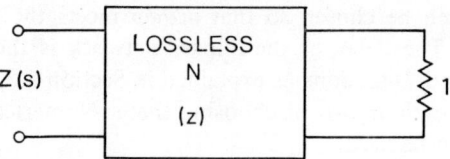

FIG. 12.26. *Realisation of a positive real driving-point impedance.*

as shown in Fig. 12.27 then if, at any particular value of s, there is no transmission through *any one* of these simpler networks, there is no transmission through the overall network. Thus if $s_1, s_2, s_3 \ldots s_k$ are the zeros of transmission of the original network and N_1 has a zero of transmission at s_1, N_2 at s_2, N_3 at $s_3, \ldots N_k$ at s_k, then the overall zeros of transmission of the cascade are the same as those of the original network. In cascade synthesis we therefore try to realise the lossless network N as a cascade of lossless sub-networks N_i, each of which possesses some of the transmission zeros of N. This has the advantage that the interconnection of the sub-networks is simple and requires no ideal transformers while individual transmission zeros are controlled by only one sub-network.

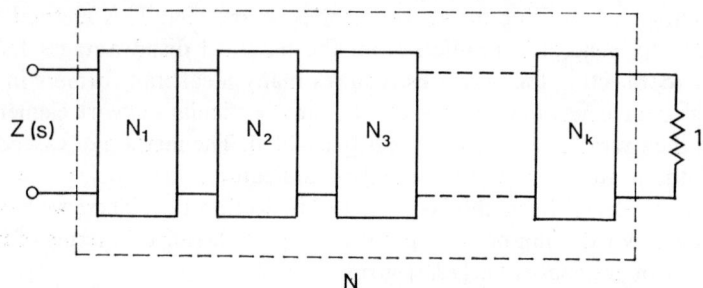

FIG. 12.27. *Realisation of the network N in Fig. 12.26 as a cascade of sub-networks.*

Let the required driving-point impedance be

$$Z(s) = \frac{m_1 + n_1}{m_2 + n_2}$$

with m even and n odd.

$$Z(s) + Z(-s) = \frac{2(m_1 m_2 - n_1 n_2)}{m_2^2 - n_2^2} \qquad [12.86]$$

TRANSFER FUNCTION SYNTHESIS

Now the zeros of transmission are the zeros of $m_1 m_2 - n_1 n_2$ and we shall assume that this is a perfect square. (If not, $Z(s)$ could be augmented as described in Chapter 11 to make it so.) We shall also assume that the Brune preamble has been carried out so that $Z(s)$ has no poles or zeros on the imaginary axis. Now since $m_1 m_2 - n_1 n_2$ is the numerator of z_{12}^2, it must be the square of an even function. We can thus write, in general,

$$m_1 m_2 - n_1 n_2 = \left[\prod_1^{k_1} \left(1 + \frac{s^2}{\omega_i^2}\right) \prod_1^{k_2} \left(1 - \frac{s^2}{\sigma_i^2}\right) \prod_1^{k_3} \{s^4 + 2(\omega_i^2 - \sigma_i^2)s^2 + (\omega_i^2 + \sigma_i^2)^2\} \right]^2 \quad [12.87]$$

thus giving three types of transmission zero

(i) $\quad s = \pm j\omega_i \quad$ a pair on the imaginary axis

(ii) $\quad s = \pm \sigma_i \quad$ a pair on the real axis

(iii) $\quad s = \pm \sigma_i \pm j\omega_i \quad$ a quadruplet of complex zeros.

Thus if we have three kinds of sub-networks (or sections, as they are usually called) each capable of realising one of the three types of factors in $m_1 m_2 - n_1 n_2$, then a suitable cascade of such sections will realise all the transmission zeros.

The problem is therefore the following: Given $Z(s)$ (and hence its transmission zeros) how do we specify (and realise) the various sections required? The basis for this is very simple. Suppose $Z(s)$ is to be realised as in Fig. 12.26, and suppose $s = s_1$ is a transmission zero which we wish to assign to N_1 as shown in Fig. 12.28. Then if N_1 realises this transmission zero it must not be a transmission zero of the remainder of the network (otherwise we would have gained no advantage). We must then choose the parameters of N_1 so that this takes place. Let the $ABCD$ parameters of N_1 be

$$\begin{pmatrix} A & B \\ C & D \end{pmatrix} = \frac{1}{F(s)} \begin{pmatrix} A_1 & B_1 \\ C_1 & D_1 \end{pmatrix} \quad [12.88]$$

where $F(s) = 0$ at $s = s_1$, and is an even function.

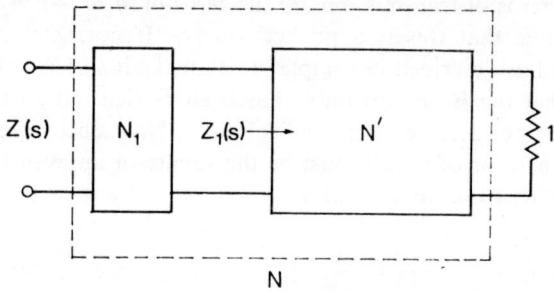

FIG. 12.28. *Realisation of the network N in Fig. 12.26 as a cascade of a sub-network N_1 and a remaining network N'. $Z_1(s)$ is the driving-point impedance at the input to N'.*

Then
$$Z(s) = \frac{A_1 Z_1(s) + B_1}{C_1 Z_1(s) + D_1}$$

so that
$$Z_1(s) = \frac{B_1 - ZD_1}{ZC_1 - A_1} \qquad [12.89]$$

and
$$Z_1(s) + Z_1(-s) = \frac{(A_1 D_1 - B_1 C_1)\{Z(s) + Z(-s)\}}{\{A_1 - C_1 Z(s)\}\{A_1 + C_1 Z(-s)\}} \qquad [12.90]$$

(remembering that since N_1 is lossless, A_1 and D_1 are even, while B_1 and C_1 are odd). Also, since N_1 is reciprocal

$$A_1 D_1 - B_1 C_1 = F^2(s) \qquad [12.91]$$

and since $s = s_1$ is a zero of transmission of $Z(s)$, $F^2(s)$ is a factor of $Z(s) + Z(-s)$. Thus
$$Z_1(s) + Z_1(-s) = \frac{F^2(s)\{Z(s) + Z(-s)\}}{\{A_1 - C_1 Z(s)\}\{A_1 + C_1 Z(-s)\}} \qquad [12.92]$$

Now what we require is that s_1 *should not* be a zero of $(Z_1(s) + Z_1(-s))$ if we are to succeed, so that factor $F^4(s)$ contained in the numerator should also be contained in the denominator. This requires that $A_1 - C_1 Z(s)$ should have a factor $F^2(s)$ since

$$A_1 + C_1 Z(-s) = \{A_1 - C_1 Z(s)\} + C_1\{Z(s) + Z(-s)\}$$

and $Z(s)+Z(-s)$ has a factor $F^2(s)$ so if $A_1 - C_1 Z(s)$ has this factor, so has $A_1 + C_1 Z(-s)$, giving an overall factor $F^4(s)$ in the denominator which cancels with the same factor in the numerator. Thus the condition on the network N_1 is that $A_1 - C_1 Z(s)$ should have a factor $F^2(s)$. The synthesis then proceeds by taking $Z_1(s)$, calculated from [12.89], and repeating the procedure for some other transmission zero, and so on until finally only the terminating resistor is left. In a rigorous treatment it would be necessary to show that $Z_1(s)$ as given in [12.89] is positive real and of lesser degree than $Z(s)$ and that the network N_1 as specified by its $ABCD$ parameters is realisable.

As already observed, we need only three types of section to realise any impedance. These are:

(i) *The Brune section*

This realises imaginary axis transmission and has

$$\begin{pmatrix} A & B \\ C & D \end{pmatrix} = \frac{1}{1+(s^2/\omega_i^2)} \begin{pmatrix} 1+as^2 & bs \\ cs & 1+ds^2 \end{pmatrix} \qquad [12.93]$$

with $(1+as^2)(1+ds^2) - bcs^2 = [1+(s^2/\omega_i^2)]^2$

(ii) *The C-section*

This realises real axis transmission zeros and has

$$\begin{pmatrix} A & B \\ C & D \end{pmatrix} = \frac{1}{1+(s^2/\sigma_i^2)} \begin{pmatrix} 1+as^2 & bs \\ cs & 1+ds^2 \end{pmatrix} \qquad [12.94]$$

with $(1+as^2)(1+ds^2) - bcs^2 = [1-(s^2/\sigma_i^2)]^2$

(iii) *The D-section*

This realises complex transmission zeros and has

$$\begin{pmatrix} A & B \\ C & D \end{pmatrix}$$
$$= \frac{1}{(\sigma_i^2+\omega_i^2)^2 + 2(\omega_i^2 - \sigma_i^2)s^2 + s^4} \begin{pmatrix} s^4 + a_1 s^2 + a_2 & b_1 s + b_2 s^3 \\ c_1 s + c_2 s^3 & s^4 + d_1 s^2 + d_2 \end{pmatrix}$$
$$[12.95]$$

with

$$(s^4 + a_1 s^2 + a_2)(s^4 + d_1 s^2 + d_2) - (b_1 s + b_2 s^3)(c_1 s + c_2 s^3)$$
$$= \{(\sigma_i^2 + \omega_i^2)^2 + 2(\omega_i^2 - \sigma_i^2)s^2 + s^4\}^2$$

Thus we choose one of the forms given in [12.93], [12.94], [12.95], as appropriate and write the condition that $A_1 - C_1 Z(s)$ has a factor $F^2(s)$ to determine A_1 and C_1. B_1 and D_1 are found from the condition that $A_1 D_1 - B_1 C_1 = F^2(s)$ so that the $ABCD$ parameters of N_1 are then known. N_1 can be synthesised by any of the methods given in section 10.2. $Z_1(s)$ is then formed and the realisation proceeds.

If $A_1 - C_1 Z(s)$ has a factor $F^2(s)$ then

$$A_1 - C_1 Z(s) = 0 \text{ for } F(s) = 0$$

and

$$(d/ds)\{A_1 - C_1 Z(s)\} = 0 \text{ for } F(s) = 0$$

If $s = s_1$ is a zero of $F(s)$ then

$$A_1(s_1) - C_1(s_1) Z(s_1) = 0$$
$$A_1'(s_1) - C_1'(s_1) Z(s_1) - C_1(s_1) Z'(s_1) = 0 \quad [12.96]$$

where prime denotes differentiation with respect to s. Thus A_1 and C_1 are functions only of the transmission zero and the impedance and its derivative evaluated at the transmission zero. Explicit formulae giving these relationships have been derived for all three cases.

As an example, consider the impedance function

$$Z(s) = \frac{s^4 + 13s^3 + 3s^2 + 10s + 1}{16s^4 + 5s^3 + 17s^2 + 2s + 1} \quad [12.97]$$

which has no j-axis poles or zeros. First we calculate the function $m_1 m_2 - n_1 n_2$ which is

$$(s^4 + 3s^2 + 1)(16s^4 + 17s^2 + 1) - (13s^3 + 10s)(5s^3 + 2s)$$
$$= 16s^8 - 8s^4 + 1$$
$$= (4s^4 - 1)^2$$

Thus $m_1 m_2 - n_1 n_2$ is a perfect square and the transmission zeros are given by

$$4s^4 - 1 = 0$$
$$(2s^2 - 1)(2s^2 + 1) = 0 \quad [12.98]$$

or

$$s = \pm 1/\sqrt{2} \quad \text{a pair of real zeros}$$
$$s = \pm j(1/\sqrt{2}) \quad \text{a pair of imaginary zeros.}$$

TRANSFER FUNCTION SYNTHESIS 527

Thus in a cascade synthesis we shall require one Brune section to realise the imaginary transmission zeros and one C-section to realise the real transmission zeros.

Let us take the Brune section first. (This is quite arbitrary as the transmission zeros may be realised in any order.) The $ABCD$ parameters of this section are given in [12.93] and we require that equations [12.96] be satisfied for $s_1 = \pm j/\sqrt{2}$. These give

$$(1+as^2) - csZ(s) = 0 \quad \text{for } s = \pm j/\sqrt{2}$$
$$2as - cZ(s) - csZ'(s) = 0 \quad \text{for } s = \pm j/\sqrt{2}$$

where $Z(s)$ is given in [12.97]. The unknown parameters are a and c and the above equations may be solved simultaneously to give

$$c = \frac{2}{s\{Z(s) - sZ'(s)\}} \quad s = \pm j/\sqrt{2}$$

$$a = \frac{Z(s) + sZ'(s)}{s^2\{Z(s) - sZ'(s)\}} \quad s = \pm j/\sqrt{2}$$

Now, in general, if

$$Z(s) = \frac{m_1 + n_1}{m_2 + n_2}$$

multiplying numerator and denominator by $m_2 - n_2$ gives

$$Z(s) = \frac{m_1 m_2 - n_1 n_2}{m_2^2 - n_2^2} + \frac{n_1 m_2 - n_2 m_1}{m_2^2 - n_2^2}$$

and evaluating at a transmission zero we need only evaluate the second term since for such values of s, $m_1 m_2 - n_1 n_2$ is zero. So we have only to evaluate an odd function. Thus $Z(s)$ evaluated at an imaginary transmission zero is always imaginary.

In the present case,

$$Z(\pm j/\sqrt{2}) = \mp j/\sqrt{2}$$

and

$$Z'(\pm j/\sqrt{2}) = 3$$

thus

$$c = \frac{2}{(\pm j/\sqrt{2})\{(\mp j/2) \mp (3j/2)\}}$$
$$= 1$$

$$a = \frac{(\mp j/\sqrt{2}) \pm (3j/\sqrt{2})}{-\frac{1}{2}\{(\mp j/\sqrt{2}) \pm (3j/\sqrt{2})\}}$$

$$= 1$$

But since $(1+as^2)(1+ds^2) - bcs^2 = (1+2s^2)^2$

$$d = 4$$

$$b = 1$$

So the Brune section has an *ABCD* matrix

$$\frac{1}{1+2s^2}\begin{pmatrix} 1+s^2 & s \\ s & 1+4s^2 \end{pmatrix} \qquad [12.99]$$

Forming the corresponding z-matrix shows that this is the same circuit as the example given in [*10.37*] and realised in Fig. 10.29. We must now form the remainder impedance according to [*12.84*]. This gives

$$Z_1(s) = \frac{4s^6 + 36s^5 + 8s^4 + 36s^3 + 5s^2 + 9s + 1}{16s^6 + 4s^5 + 20s^4 + 4s^3 + 8s^2 + s + 1}$$

$$= \frac{(2s^2+1)^2(s^2+9s+1)}{(2s^2+1)^2(4s^2+s+1)}$$

$$= \frac{s^2+9s+1}{4s^2+s+1} \qquad [12.100]$$

$Z_1(s)$ when formed in this way always has a factor which is the square of the factor containing the transmission zeros, in both numerator, and denominator, thus ensuring that it is of lower degree than $Z(s)$ after cancelling this factor.

We now proceed with $Z_1(s)$ as given by [*12.100*]. First we know that it should possess the transmission zeros at $s = \pm 1\sqrt{2}$. If we form the appropriate value of $m_1 m_2 - n_1 n_2$ we obtain

$$(s^2+1)(4s^2+1) - 9s^2 = (1-2s^2)^2$$

as we expect. This will require a *C*-section whose *ABCD* matrix is given by [*12.94*] and we require

$$1 + as^2 - csZ_1(s) = 0 \qquad \text{for } s = \pm 1/\sqrt{2}$$

$$2as - cZ_1(s) - csZ_1'(s) = 0 \qquad \text{for } s = \pm 1/\sqrt{2}$$

TRANSFER FUNCTION SYNTHESIS

These give the same equations as before for a and c (s now being $\pm 1/\sqrt{2}$ instead of $\pm j/\sqrt{2}$). In this case

$$Z(s) = \pm 3/\sqrt{2}$$

and

$$Z'(s) = -1$$

Hence

$$c = 1$$
$$a = 1$$

and

$$(1+s^2)(1+ds^2) - bs^2 = (1-2s^2)^2$$

so

$$d = 4$$
$$b = 9$$

Thus the C-section has $ABCD$ matrix

$$\frac{1}{1-2s^2}\begin{pmatrix} 1+s^2 & 9s \\ s & 1+4s^2 \end{pmatrix} \qquad [12.101]$$

The corresponding z-parameters are

$$z_{11} = \frac{1+s^2}{s}; \quad z_{12} = \frac{1-2s^2}{s}; \quad z_{22} = \frac{1+4s^2}{s}$$

These can readily be realised by the methods of section 10.2 to give the circuit shown in Fig. 12.29. The remaining impedance is, from [12.89]

$$Z_2(s) = \frac{4s^4 - 4s^2 + 1}{4s^4 - 4s^2 + 1}$$

$$= \frac{(2s^2-1)^2}{(2s^2-1)^2}$$

$$= 1$$

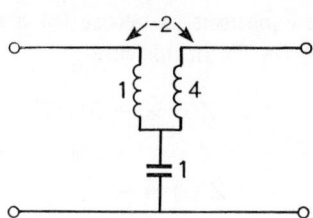

FIG. 12.29. *Realisation of the ABCD matrix given in* [*12.101*].

so the terminating impedance is a one ohm resistor. The complete realisation is shown in Fig. 12.30.

Although the above example involves some rather tedious working it is possible in practice to avoid most of this using the explicit formulae (for details see J. O. Scanlan and J. D. Rhodes, 1970. Unified Theory of Cascade Synthesis. *Proc. I.E.E.*, *117*, 665). We may also consider the process of cascade synthesis from the point of view of transfer function synthesis. The zeros of transmission are the zeros of $S_{12}(p)S_{12}(-p)$, which from [*12.28*] are either zeros of $Z_{in}(p)+Z_{in}(-p)$ or values of p which are simultaneously poles of $Z_{in}(p)$ and $Z_{in}(-p)$. The latter can only occur at imaginary values of p. In the subsequent synthesis of $Z_{in}(p)$ such poles are removed in the Brune preamble and the procedure for cascade synthesis is therefore that already outlined.

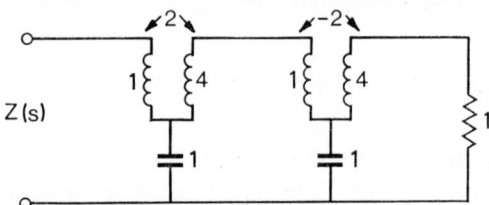

FIG. 12.30. *Realisation of the driving-point impedance given in* [*12.97*].

In the special case of an allpass network, described in Chapter 8, S_{12} is of the form

$$S_{12}(p) = \frac{Q(-p)}{Q(p)} \qquad [12.102]$$

TRANSFER FUNCTION SYNTHESIS

and $Z_{in}(p) = R$ for all values of p. However, the network does in a sense have transmission zeros given by $Q(p)Q(-p) = 0$. Thus S_{12} can be realised by cascade synthesis. If the realisation is as shown in Fig. 12.31, then the lossless network N is described by

$$\frac{1}{F(s)}\begin{pmatrix} \hat{A} & \hat{B} \\ \hat{C} & \hat{D} \end{pmatrix}$$

and

$$Z_{in} = \frac{\hat{A}R + \hat{B}}{\hat{C}R + \hat{D}} = R$$

or

$$(\hat{A} - \hat{D})R + \hat{B} = \hat{C}R^2$$

for all values of p.

Hence

$$\hat{A} = \hat{D}$$

$$\hat{B} = \hat{C}R^2$$

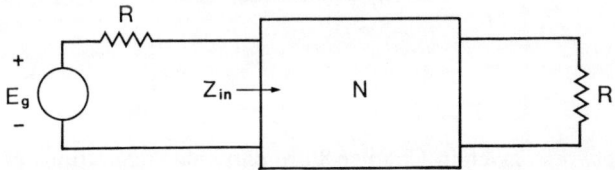

FIG. 12.31. *Realisation of an allpass network.*

So N is symmetrical and has a transmission matrix

$$\frac{1}{F(s)}\begin{pmatrix} \hat{A} & \hat{C}R^2 \\ \hat{C} & \hat{A} \end{pmatrix}$$

If this is then realised as a cascade of sections each having some of the transmission zeros then each section must be an allpass and so must have $A = D$, $B = CR^2$. In the case of a C-section we have therefore

$$\frac{1}{1-(s^2/\sigma_i^2)}\begin{pmatrix} 1+as^2 & cR^2s \\ cs & 1+as^2 \end{pmatrix}$$

and

$$(1+as^2)^2 - c^2R^2s^2 = [1-(s^2/\sigma_i^2)]^2$$

J

so that
$$a = 1/\sigma_i^2$$
$$c = 2/R\sigma_i$$

So an allpass C-section has

$$\begin{pmatrix} A & B \\ C & D \end{pmatrix} = \frac{1}{1-(s^2/\sigma_i^2)} \begin{pmatrix} 1+(s^2/\sigma_i^2) & (2R/\sigma_i)s \\ (2/R\sigma_i)s & 1+(s^2/\sigma_i^2) \end{pmatrix} \quad [12.103]$$

In the case of an allpass D-section a similar analysis shows that in [12.95] the parameters are

$$a_1 = 2(\omega_i^2 + 3\sigma_i^2) = d_1$$
$$a_2 = (\sigma_i^2 + \omega_i^2)^2 = d_2$$
$$b_1 = R^2 c_1$$
$$b_2 = R^2 c_2 \qquad [12.104]$$
$$c_1 = \frac{4\sigma_i(\sigma_i^2 + \omega_i^2)}{R}$$
$$c_2 = \frac{4\sigma_i}{R}$$

The networks given in Chapter 8 are particular realisations of these parameters.

We have thus seen, in general terms, how the techniques of cascade synthesis may be used for the realisation of driving point impedances and transfer functions, with some resulting advantages. The methods described also have considerable importance in theoretical investigations of many novel synthesis procedures.

12.6. Principles of Broadband Matching Theory

In all problems of transfer function synthesis we have considered so far, it has been assumed that the specified terminating impedances were purely resistive. It is obviously of interest to consider the problem of transfer function realisation when the termination is a specified impedance which is not necessarily real. Such situations can arise, for example, if the termination is the input of some active device whose input impedance has both a real and an imaginary part, or if the termination is an antenna or a transducer where again the input impedance is not purely real. We shall consider the situation shown in Fig. 12.32. in which $Z_L(p)$ is a positive real terminating impedance and we are interested in the ratio of the power delivered to Z_L to the available power from the generator (whose impedance is assumed to be purely real in order to simplify the discussion).

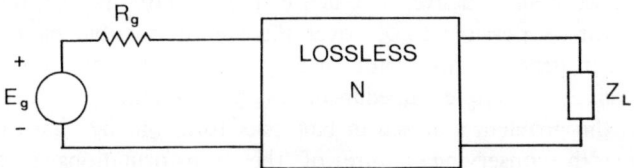

FIG. 12.32. *The general broadband matching problem.*

The question we ask is: what constraints are placed on the transducer power gain by Z_L? Now since $Z_L(p)$ is positive real we could synthesise it by the Darlington method in the form of a lossless network N_1 terminated in a one ohm resistor. (Although it should be emphasised that this may not correspond to the physical circuit or device which gives rise to Z_L.) The resulting circuit is shown in Fig. 12.33. Thus the overall circuit consists of a lossless network

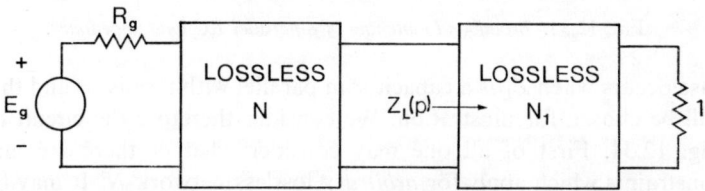

FIG. 12.33. *An alternative realisation of the circuit in Fig. 12.32 where $Z_L(p)$ has been realised by the Darlington procedure.*

(N and N_1 in cascade), terminated in a one ohm resistor. Since N_1 is lossless the power delivered to $Z_L(p)$ is the power delivered to the one ohm resistor, and since N_1 is fixed by Z_L we see that the overall lossless network is thereby constrained. The elucidation of these constraints in general is beyond the scope of this book, but some observations can be made. Clearly, in most cases, the transmission zeros of N_1 are zeros of the overall transfer function since N and N_1 are in the cascade. One can then develop conditions on the overall transducer power gain in terms of its behaviour at the transmission zeros of N_1 which is constrained by the prescribed form of Z_L. (See D. C. Youla, 1964, A new theory of broadband matching. *I.E.E.E. Trans. on Circuit Theory*, CT-11, 30.)

Apart, however, from the general theoretical results such as those mentioned above the ideas have direct application in a variety of practical problems. A situation frequently encountered in practice is that we wish to deliver as much of the available power from the generator as possible to Z_L over the widest possible bandwidth. Again, in general, this is not an easy problem to solve but, if Z_L is a relatively simple impedance, results can readily be deduced. Often the problem is posed in bandpass form but by virtue of the bandwidth conserving nature of the transformations given in Chapter 8 we need only consider a lowpass version. The simplest

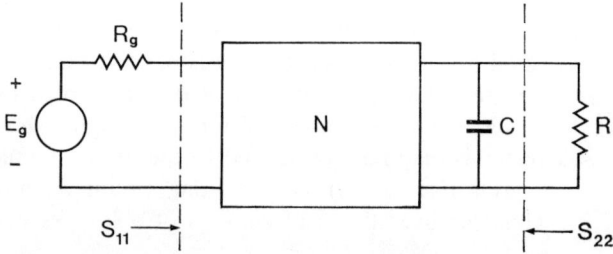

FIG. 12.34. *Broadband matching of a parallel RC load impedance.*

case occurs when Z_L is a capacitor in parallel with a resistor and this will be chosen for illustration. We consider therefore the circuit of Fig. 12.34. First of all one may consider whether there are any constraints which apply for *arbitrary* lossless network N. It may be shown that whatever the exact form of N

$$\int_0^\infty \ln \frac{1}{|S_{22}(j\omega)|} d\omega \leq \frac{\pi}{RC} \qquad [12.105]$$

but since N is lossless

$$|S_{22}(j\omega)|^2 = |S_{11}(j\omega)|^2 = 1 - |S_{12}(j\omega)|^2$$

so

$$\int_0^\infty \ln \frac{1}{|S_{11}(j\omega)|} d\omega \leq \frac{\pi}{RC} \qquad [12.106]$$

TRANSFER FUNCTION SYNTHESIS

The shunt capacitor C requires that the admittance, y, corresponding to S_{22} has a pole at infinity with residue at least as large as C. This in turn implies that for p sufficiently large, and ignoring terms in $1/p^2$ and higher powers,

$$S_{22}(p) \to -\left(1 + \frac{2}{RCp}\right)$$

The Cauchy Integral Theorem may then be applied to $ln(1/S_{22})$ using the imaginary axis and the semicircle at infinity as the contour to yield [12.105]. The presence of right half plane zeros in S_{22} reduces the right hand side of [12.105] so equality can only be achieved if the zeros of S_{22} are entirely in the left half plane.

The form of [12.106] clearly shows that the area under a curve of $\ln(1/|S_{11}|)$ as a function of frequency is constrained. Good matching of power transfer to the load occurs at those frequencies where $|S_{11}|$ is small while regions where $|S_{11}| \to 1$ contribute least to the integral. Thus for matching purposes we should arrange that $|S_{11}| = 1$ (with $|S_{12}| = 0$) outside the band of interest $(0 - \omega_c)$, in that case [12.106] becomes

$$\int_0^{\omega_c} \ln \frac{1}{|S_{11}(j\omega)|} d\omega \leq \frac{\pi}{RC} \qquad [12.107]$$

Within the passband it is clear that a decrease in $|S_{11}|$ in one region can only occur at the expense of an increase in another region and since we are interested in keeping the minimum value of $|S_{12}|$ over the band as large as possible the optimum form of $|S_{11}|$ (and $|S_{12}|$) within the passband is a constant. Figure 12.35 shows the optimum transducer power gain. In this case

$$|S_{11}|^2 = 1 - T \qquad |\omega| \leq |\omega_c|$$

and [12.107] becomes

$$\omega_c \ln\left\{\frac{1}{(1-T)^{\frac{1}{2}}}\right\} \leq \frac{\pi}{RC} \qquad [12.108]$$

Although [12.108] refers to the idealised response of Fig. 12.35 which requires N to be infinite, its simple form, and the fact that it is the best possible performance, enable one to rapidly establish

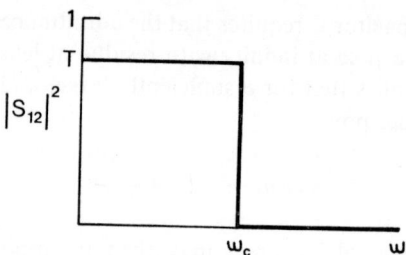

FIG. 12.35. *Optimum transducer power gain for broadband matching of a parallel RC load.*

whether a particular specification on T and ω_c is likely to be possible or not.

As a more practical proposition suppose we try to choose N so that $|S_{12}|^2$ is maximally flat. The constraints could be found by using [12.105] but this is rather cumbersome and a more convenient approach is to use the explicit formulae for the element values. The constraint is that the final element adjacent to the load resistor should be a shunt capacitor whose value is greater than or equal to C. Now we shall require a slightly different form of $|S_{12}|^2$ than was used previously since we do not want $|S_{12}(0)| = 1$ with, correspondingly, $|S_{11}(0)| = 0$ and consequent reduction in bandwidth by virtue of [12.105]. Thus we choose

$$|S_{12}|^2 = \frac{k}{1+(\alpha\omega)^{2n}} \qquad [12.109]$$

Where α is a bandwidth scaling factor. Thus

$$|S_{11}|^2 = \frac{1-k+(\alpha\omega)^{2n}}{1+(\alpha\omega)^{2n}} \qquad [12.110]$$

and it is convenient to put

$$K^{2n} = 1-k$$

The recurrence formulae for the element values are then found to be

$$g_1 = \frac{2\alpha \sin(\pi/2n)}{1-K}$$

$$g_r g_{r+1} = \frac{4\alpha^2 \sin\{(2r-1)\pi/2n\} \sin\{(2r+1)\pi/2n\}}{1-2K\cos(r\pi/n)+K^2}$$

$$[12.111]$$

TRANSFER FUNCTION SYNTHESIS

We may in fact take g_1 to be the capacitance at the load end since turning the overall lossless network end for end does not affect $|S_{12}|^2$. The formulae in [12.48] are obtained from [12.111] by setting $K = 0, \alpha = 1$. In [12.111] it is implicit that the terminating resistance is 1 ohm, and α is the reciprocal of the frequency at which $|S_{12}|^2 = 0.5k$. The constraint imposed by the capacitor C is that taking the first element to be a shunt capacitor it must not be less than C when the terminating resistor is R or not less than RC for a one ohm terminating resistor. Thus

$$g_1 \geqq RC \qquad [12.112]$$

or

$$\frac{2\alpha \sin(\pi/2n)}{1-K} \geqq RC \qquad [12.113]$$

Now if ω_c is the frequency at which $|S_{12}|^2 = T$, the permitted tolerance, then

$$T = \frac{k}{1+(\alpha\omega_c)^{2n}} \qquad [12.114]$$

and we find

$$RC\omega_c \leqq \frac{2(k-T)^{1/2n}}{T^{1/2n}\{1-(1-k)^{1/2n}\}} \sin(\pi/2n) \qquad [12.115]$$

and we must have

$$1 \geqq k \geqq T \qquad [12.116]$$

Now, for a given T we must choose k to optimise ω_c in [12.115]. If $k = T$ clearly the right hand side is zero, and if $k = 1$ it is

$$2[(1/T)-1]^{1/2n} \sin(\pi/2n)$$

To find the optimum we set

$$\frac{d}{dk}\left[\frac{(k-T)^{1/2n}}{1-(1-k)^{1/2n}}\right] = 0$$

This gives

$$k = 1-(1-T)^{2n/2n-1} \qquad [12.117]$$

This value of k ensures a maximum bandwidth for a given value of T. In accordance with [12.116] it is necessary that $k \geqq T$. But

$$k-T = (1-T)-(1-T)^{2n/2n-1}$$

and since $1-T < 1$ this is positive.

Substituting the optimum value of k from [12.117] in [12.115] we obtain

$$RC\omega_c \leqq \frac{2(1-T)^{1/2n} \sin(\pi/2n)}{T^{1/2n}\{1-(1-T)^{1/(2n-1)}\}^{(2n-1)/2n}} \qquad [12.118]$$

and comparing this with the absolute maximum bandwidth given in [12.108] we get

$$RC\omega_c \leqq \frac{\pi}{\ln(1-T)^{-\frac{1}{2}}} \qquad [12.119]$$

In the case of a Chebyshev response we again require

$$g_1 \geqq RC$$

where now g_1 has the value appropriate to this type of response. As for the maximally flat case we take g_1 to be the element nearest the load and require it to be a shunt capacitor. In this case we choose

$$|S_{12}|^2 = \frac{1}{1+h^2+\varepsilon^2 T_n^2(\omega/\omega_c)} \qquad [12.120]$$

Where the permitted tolerance T is $1/(1+h^2+\varepsilon^2)$ and ω_c is the corresponding bandwidth. For a response of the type given in [12.120] the recurrence formulae for the element values become

$$\omega_c g_1 = \frac{2\sin(\pi/2n)}{x-y}$$

$$\omega_c^2 g_r g_{r+1} = \frac{4\sin\{(2r-1)\pi/2n\}\sin\{(2r+1)\pi/2n\}}{x^2+y^2+\sin^2(r\pi/n)-2xy\cos(r\pi/n)} \qquad [12.121]$$

$$r = 1, 2, 3 \ldots n-1$$

With

$$x = \sinh\left\{\frac{1}{n}\sinh^{-1}\left(\frac{1+h^2}{\varepsilon^2}\right)^{\frac{1}{2}}\right\}$$

$$y = \sinh\left\{\frac{1}{n}\sinh^{-1}\frac{h}{\varepsilon}\right\}$$

and we require

$$RC\omega_c \leqq \frac{2\sin(\pi/2n)}{x-y} \qquad [12.122]$$

TRANSFER FUNCTION SYNTHESIS

For maximum bandwidth with a given tolerance T, $x-y$ must be minimised. Now

$$T = \frac{1}{1+h^2+\varepsilon^2}$$

or

$$h = \left\{\frac{1}{T}-1-\varepsilon^2\right\}^{\frac{1}{2}}$$

Thus we must choose the ripple level, ε, to minimise

$$\sinh\left\{\frac{1}{n}\sinh^{-1}\left(\frac{1}{\varepsilon^2 T}-1\right)^{\frac{1}{2}}\right\} - \sinh\left\{\frac{1}{n}\sinh^{-1}\left(\frac{1-T}{\varepsilon^2 T}-1\right)^{\frac{1}{2}}\right\}$$

with respect to variation in ε. Setting the derivatives equal to zero gives

$$\left(\frac{1-T}{T\varepsilon^2}-1\right)^{\frac{1}{2}} \cosh\left\{\frac{1}{n}\sinh^{-1}\left(\frac{1}{T\varepsilon^2}-1\right)^{\frac{1}{2}}\right\}$$
$$= (1-T)^{\frac{1}{2}}\left(\frac{1}{T\varepsilon^2}-1\right)^{\frac{1}{2}} \cosh\left\{\frac{1}{n}\sinh^{-1}\left(\frac{1-T}{T\varepsilon^2}-1\right)^{\frac{1}{2}}\right\} \quad [12.123]$$

But since $\sinh^{-1}\theta = \cosh^{-1}(1+\theta^2)^{\frac{1}{2}}$, [12.123] may also be written as

$$\left(\frac{1-T}{T\varepsilon^2}-1\right)^{\frac{1}{2}} \cosh\left\{\frac{1}{n}\cosh^{-1}\left(\frac{1}{T\varepsilon^2}\right)^{\frac{1}{2}}\right\}$$
$$= (1-T)^{\frac{1}{2}}\left(\frac{1}{T\varepsilon^2}-1\right)^{\frac{1}{2}} \cosh\left\{\frac{1}{n}\cosh^{-1}\left(\frac{1-T}{T\varepsilon^2}\right)^{\frac{1}{2}}\right\} \quad [12.124]$$

This equation can be written in polynomial form but this does not lead to an explicit solution for ε. Instead we put

$$K = \cosh\left\{\frac{1}{n}\cosh^{-1}\left(\frac{1}{T\varepsilon^2}\right)^{\frac{1}{2}}\right\} \quad [12.125]$$

or

$$\left(\frac{1}{T\varepsilon^2}\right)^{\frac{1}{2}} = \cosh(n\cosh^{-1}K) = T_n(K) \quad [12.126]$$

where $T_n(K)$ is the Chebyshev function of degree n and argument K. [12.124] may then be rewritten as

$$K\{(1-T)T_n^2(K)-1\}^{\frac{1}{2}} = (1-T)^{\frac{1}{2}}(T_n^2(K)-1)^{\frac{1}{2}}$$
$$\times \cosh\left[(1/n)\cosh^{-1}\{(1-T)^{\frac{1}{2}}T_n(K)\}\right]$$

or

$$\cosh\left[(1/n)\cosh^{-1}\{(1-T)^{\frac{1}{2}}T_n(K)\}\right] = K\left\{\frac{T_n^2(K)-1/(1-T)}{T_n^2(K)-1}\right\}^{\frac{1}{2}}$$

[12.127]

Which can also be written as

$$(1-T)^{\frac{1}{2}}T_n(K) = \cosh\left[n\cosh^{-1}\left\{K\left(\frac{T_n^2(K)-1/(1-T)}{T_n^2(K)-1}\right)^{\frac{1}{2}}\right\}\right]$$

or

$$(1-T)^{\frac{1}{2}}T_n(K) = T_n\left[K\left\{\frac{T_n^2(K)-1/(1-T)}{T_n^2(K)-1}\right\}^{\frac{1}{2}}\right]$$ [12.128]

From [12.126] $T_n(K) > 1$ so $K > 1$ and since

$$h^2 + \varepsilon^2 = (1/T) - 1$$
$$\varepsilon^2 \leq (1/T) - 1$$

or

$$1/(\varepsilon^2 T) \geq 1/(1-T)$$

so

$$T_n^2(K) \geq 1/(1-T) \qquad [12.129]$$

Thus the argument of the Chebyshev function on the right hand side of [12.128] is positive and, since [12.129] ensures that the left hand side is greater than unity, [12.128] requires this argument to be greater than unity.

By expanding the Chebyshev functions in [12.128] a polynomial in K may be formed and solved (numerically if necessary). Substituting this result in [12.122] gives

$$RC\omega_c \leq \frac{2\sin(\pi/2n)}{x-y} \qquad [12.130]$$

with

$$x-y = \sinh\left\{\frac{1}{n}\sinh^{-1}\left(\frac{1}{\varepsilon^2 T}-1\right)^{\frac{1}{2}}\right\} - \sinh\left\{\frac{1}{n}\sinh^{-1}\left(\frac{1-T}{\varepsilon^2 T}-1\right)^{\frac{1}{2}}\right\}$$

$$= \sinh\left\{\frac{1}{n}\cosh^{-1}\left(\frac{1}{\varepsilon^2 T}\right)^{\frac{1}{2}}\right\} - \sinh\left\{\frac{1}{n}\cosh^{-1}\left(\frac{1-T}{\varepsilon^2 T}\right)^{\frac{1}{2}}\right\}$$

$$= \{\cosh^2(\cosh^{-1}K) - 1\}^{\frac{1}{2}} - \left[\cosh^2\left\{\frac{1}{n}\cosh^{-1}\left(\frac{1-T}{\varepsilon^2 T}\right)^{\frac{1}{2}}\right\} - 1\right]^{\frac{1}{2}}$$

which from [*12.127*] gives

$$x-y = (K^2-1)^{\frac{1}{2}} - \left[K^2\left\{\frac{T_n^2(K) - 1/(1-T)}{T_n^2(K) - 1}\right\} - 1\right]^{\frac{1}{2}} \qquad [12.131]$$

The procedure is therefore to solve [*12.128*] for K, evaluate [*12.131*] and substitute in [*12.130*] to find the optimum constraint on the bandwidth.

We have thus seen how, in principle, a generator can be matched to a parallel RC load using an infinite matching network or one which realises a maximally flat or Chebyshev response. It is found in practice that the bandwidth obtainable using a Chebyshev response of modest degree (say 5 or 6) is only slightly less than that obtained with an infinite network with the same tolerance. A notable point in all the responses is that in the optimum case a perfect match is not achieved at any frequency and the common practical method of having a perfect match at band centre is not optimum if maximum bandwidth for a prescribed tolerance is required.

Having considered the simplest case of a parallel RC load the next most difficult problem is to consider a load of the form shown in Fig. 12.36. This leads to two integral restrictions in the general case,

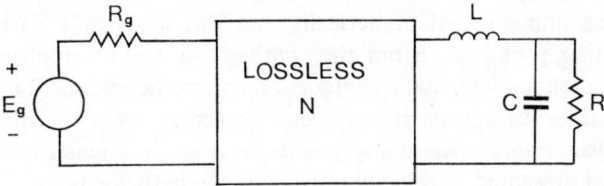

FIG. 12.36. *Broadband matching of a load impedance with two transmission zeros at infinity.*

and requires $g_1 = RC$ and $g_2 \geq L/R$ in the maximally flat and equipment cases. The general integral restrictions are

$$\int_0^\infty \ln(1/|S_{11}(j\omega)|)d\omega \leq \frac{\pi}{RC}$$

$$\int_0^\infty \omega^2 \ln(1/|S_{11}(j\omega)|)d\omega \leq \frac{\pi}{3}\left(\frac{3R^2C-L}{LR^3C^3}\right) \qquad [12.132]$$

and the more severe of these limits the bandwidth. The first restriction must be satisfied with the equality sign if no right half plane zeros of S_{22} are allowed, since C is inaccessible.

Any more complicated load may be treated by similar but rather more cumbersome procedures. The procedures can also be modified to determine the effect of parasitic reactances in active devices and are, in general, extremely versatile and valuable tools in many applications. In particular, the use of the integral constraints enables ultimate performance limits to be determined independent of any network within the designer's control.

12.7. General Observations

In these two volumes an attempt has been made to present the fundamentals of network analysis and synthesis. In the material on analysis only lumped-element, passive, linear networks have been considered specifically but the extension to active networks is straightforward while the extension to nonlinear networks is difficult and not many of the real problems have yet been fully solved.

Many analysis procedures, although straightforward in concept, must be implemented numerically and this leads to a variety of interesting problems. From the synthesis viewpoint emphasis has been laid on fundamental techniques and properties of functions and a rather rigorous approach has been adopted in order to emphasise the crucial importance of study in depth in this particular area. The material presented in this volume forms the basis for tackling more advanced topics and can be used to derive powerful synthesis techniques for other classes of networks such as distributed networks

TRANSFER FUNCTION SYNTHESIS

and digital networks. What has been presented is only a foundation for further study but the underlying philosophy and techniques have general application throughout the field.

Problems

12.1. Realise the following open-circuit voltage transfer functions in the form of a lossless network terminated in a resistor, without the use of ideal transformers or coupled coils.

(a) $Z_{21} = \dfrac{24}{24+18s+2s^2+s^3}$

(b) $Z_{21} = \dfrac{s^3}{s^3+6s^2+12s+24}$

(c) $Z_{21} = \dfrac{s^2}{s^3+6s^2+12s+24}$

(d) $Z_{21} = \dfrac{s^3+s}{s^3+6s^2+12s+24}$

12.2. Realise the following input reflection coefficients in the form of a ladder network terminated in a one ohm resistor.

(a) $S_{11} = \dfrac{2s^4+s^3+2s^2+s}{2s^4+3s^3+4s^2+3s+1}$

(b) $S_{11} = \dfrac{-2s^4+s^3-2s^2+s}{2s^4+3s^3+4s^2+3s+1}$

(c) $S_{11} = \dfrac{-2s^4-s^3-2s^2-s}{2s^4+3s^3+4s^2+3s+1}$

(d) $|S_{11}(j\omega)|^2 = \dfrac{1}{1+\omega^6}$

12.3. Realise the following transducer power gain functions in the form of resistively terminated lossless networks with a one ohm generator resistance, using direct synthesis methods and check the results using the recurrence formulae

(a) $|S_{12}|^2 = \dfrac{\omega^6}{1+\omega^6}$

(b) $|S_{12}|^2 = \dfrac{1}{1+\omega^{10}}$

(c) $|S_{12}|^2 = \dfrac{7/8}{1+\omega^6}$

12.4. Derive a maximally flat transducer power gain of minimum degree to meet the following specifications:

(a) $S_{12}(0) = 1$; $|S_{12}|^2 = 0\cdot 5$ at $\omega = 1$; $|S_{12}|^2 \leq 0\cdot 04$ at $\omega = 1\cdot 5$

(b) $S_{12}(0) = 1$; $|S_{12}|^2 = 0\cdot 5$ at 3 kHz; $|S_{12}|^2 \leq 0.04$ at 4·5 kHz

(c) $R_g = 1\,\Omega$; $R_L = 0\cdot 5\,\Omega$; $|S_{12}|^2 = \tfrac{1}{2}|S_{12}(0)|^2$ at $\omega = 1$; $|S_{12}|^2 \leq 0\cdot 04$ at $\omega = 1\cdot 5$

(d) $S_{12}(0) = 1$; $|S_{12}|^2 \geq 0\cdot 8$ at $\omega = 1$; $|S_{12}|^2 \leq 0\cdot 04$ at $\omega = 1\cdot 5$

12.5. Realise the following transducer power gain functions in the form of resistively terminated lossless networks with a one ohm generator resistance using direct synthesis methods and check the results using the recurrence formulae

(a) $|S_{12}|^2 = \dfrac{1}{1+0\cdot 01 T_3^2(\omega)}$

(b) $|S_{12}|^2 = \dfrac{1}{1+0\cdot 01 T_4^2(\omega)}$

12.6. Find the Chebyshev transducer power gain of minimum degree to meet the specification given in problem 12.4(d).

TRANSFER FUNCTION SYNTHESIS

12.7. Design a filter with Chebyshev response to meet the following bandpass specification.

$|S_{12}|^2 \geq 1/1\cdot01$ 1 kHz $\leq f \leq$ 1·96 kHz.

$|S_{12}|^2 \leq 0\cdot04$ $f \leq$ 0·6 kHz and $f \geq$ 3·25 kHz.

The generator resistance is 50 ohms.

12.8. In the case of the third order maximally flat delay approximation of [*12.83*], with $a = 1$,

(a) Calculate the value of ω for which $|S_{12}|^2 = 0\cdot5$

(b) Calculate the delay variation between zero and this value of ω

(c) For a maximally flat transducer power gain of degree 3 and the same 3 db bandwidth calculate the delay variation between zero and the 3 db frequency as a multiple of that calculated under (b)

(d) Compare the values of $|S_{12}|^2$ at a value of ω which is twice the 3 db frequency.

12.9. $S_{12}(p)$ represents a lowpass transfer function on which the lowpass to bandpass transformation $p \to p + (\omega_0^2/p)$ is performed. If $T_g(p)$ is the group delay characteristic of $S_{12}(p)$ derive an expression for the group delay characteristic of the resulting bandpass filter.

12.10. Realise the following functions using cascade synthesis.

(a) $Z_{in} = \dfrac{1 + 10s + 12s^2 + 80s^3 + 16s^4}{1 + 4s + 28s^2 + 8s^3 + 16s^4}$

(b) $Z_{in} = \dfrac{1 + 5s + 3s^2 + 10s^3 + s^4}{1 + s + 6s^2 + s^3 + 4s^4}$

(c) $S_{12}(p) = \dfrac{4 - 6s + 4s^2 - s^3}{4 + 6s + 4s^2 + s^3}$

(1 ohm terminations)

12.11. A load consists of a parallel *RC* combination which it is desired to match to a resistive generator over a band. Compare the bandwidths obtainable using the following schemes.

(a) Direct connection of the generator to the load, with equal load and generator resistances

(b) Insertion of an infinite matching network

(c) Insertion of a matching network with optimum maximally flat transducer power gain of degree 3

(d) Insertion of a matching network with optimum Chebyshev response of degree 3.

12.12. A load consists of an *LCR* combination as shown in Fig. 12.36, and the equality signs apply in [*12.132*] (S_{22} has no right half plane zeros)

(a) If the response is of the type shown in Fig. 12.35 calculate the resulting bandwidth and tolerance

(b) For a response with $|S_{12}|^2 = T$ for $\omega_1 \leq \omega \leq \omega_2$ and $|S_{12}|^2 = 0$ elsewhere show that the resulting bandwidth $(\omega_2 - \omega_1)$ and tolerance restrictions are the same as in (a) if L is replaced by $L/(1 - LC\omega_0^2)$ where $\omega_0^2 = \omega_1 \omega_2$.

APPENDIX

In this appendix we gather together for reference some mathematical results from the theory of functions of a complex variable. No attempt is made to provide rigorous mathematical proofs and for a more complete treatment the reader is referred to any standard mathematical text, e.g., E. A. Guillemin, 1950. *The Mathematics of Circuit Analysis*. The Technology Press, John Wiley and Sons Inc., New York.

(i) Analytic Functions

Let s be a complex variable $= \sigma + j\omega$. Then a rational function $F(s)$ is analytic everywhere except at its poles.

(ii) Cauchy–Riemann Equations

If $F(s) = u + jv$ then

$$\frac{\partial u}{\partial \sigma} = \frac{\partial v}{\partial \omega}$$

$$\frac{\partial u}{\partial \omega} = -\frac{\partial v}{\partial \sigma} \qquad [A\text{-}1]$$

at all analytic points.

(iii) Cauchy's Integral Formula

If $F(s)$ is analytic within a closed contour C, then

$$\oint_C F(s)\,ds = 0 \qquad [A\text{-}2]$$

and

$$F(s_0) = \frac{1}{2\pi j} \oint_C \frac{F(s)}{s - s_0}\,ds \qquad [A\text{-}3]$$

if s_0 lies within C. [A-3] is Cauchy's Integral Formula.

(iv) Maximum Modulus Theorem

If $F(s)$ is analytic in a domain D and along its boundary, then the maximum value of $|F(s)|$ lies on the boundary.

(v) Minimum Real Part Theorem

If $F(s)$ is analytic in a domain D, then the minimum value of $\text{Re } F(s)$ in D lies on the boundary.

(vi) Rouché's Theorem

If $A(s)$ and $B(s)$ are analytic in a domain D, continuous on the boundary C and $|A(s)| \leq |B(s)|$ on C then $F(s) = A(s) + B(s)$ has the same number of zeros interior to C as does $B(s)$.

SOLUTIONS TO PROBLEMS

Chapter 7

7.1. (a) Zeros at $-0.55\pm j0.835$, poles at $-0.05\pm j0.999$, peak value of $|Y|$ approximately 16.8.

(b) Two zeros coincide at -1, poles at $-0.5\pm j0.865$, peak value of $|Y|$ approximately 3.

Circuit (a) gives more rejection than (b) at $\omega = 1$.

7.2. $Z_{11}(s) = \dfrac{s^2+1}{s(s^2+2)}$

7.3. $Z_{11}(s) = \dfrac{s^2+\frac{1}{8}s+1}{s(s^2+\frac{1}{4}s+2)}$

The resistance moves the zeros at $s = \pm j$ and the poles at $s = \pm 2j$ off the j-axis slightly into the left half plane. Without the resistance $Z_{11}(s)$ is a lossless network, known as a reactance function, which always has all its poles and zeros located on the j-axis.

7.4. $Z_{11} = 1$

$Z_{12} = \dfrac{1}{s^2+\sqrt{2}s+1}$

$|Z_{12}(j\omega)|^2 = \dfrac{1}{1+\omega^4}$

$|Z_{13}(j\omega)|^2 = \dfrac{\omega^4}{1+\omega^4}$

$Z_{13} = \dfrac{s^2}{s^2+\sqrt{2}s+1}$

Since $Z_{11} = 1$ at all frequencies all power entering the network must be dissipated in the resistors at ports 2 and 3. The sum of the power dissipated in these two unity resistors is

$$i_2^2 + i_3^2 = i_1^2|Z_{12}(j\omega)|^2 + i_1^2|Z_{13}(j\omega)|^2 = i_1^2$$

as required.

7.6. $Z_{11}(s) = \dfrac{(s+1)(s+9/4)}{s+2}$

$Z_{12}(s) = \dfrac{9}{4(s+2)}$

7.7. (a) The nodal admittance matrix is

$$\begin{bmatrix} y_{11} & -y_{12} & 0 & -y_{14} \\ -y_{12} & y_{11} & -y_{14} & 0 \\ 0 & -y_{14} & y_{11} & -y_{12} \\ -y_{14} & 0 & -y_{12} & y_{11} \end{bmatrix}$$

where

$$y_{11} = s(C+C_1) + 1/sL$$

$$y_{12} = 1/sL$$

$$y_{14} = sC_1$$

The cofactors and determinant of the matrix are

$$\Delta_{11} = y_{11}(y_{11}^2 - y_{12}^2 - y_{14}^2)$$

$$\Delta_{21} = y_{12}(y_{11}^2 - y_{12}^2 + y_{14}^2)$$

$$\Delta_{31} = 2y_{11}y_{12}y_{14}$$

$$\Delta_{41} = y_{14}(y_{11}^2 + y_{12}^2 - y_{14}^2)$$

$$\Delta = (y_{11}^2 - y_{12}^2 - y_{14}^2)^2 - 4y_{12}^2 y_{14}^2$$

(b) $Z_{11} = \dfrac{\Delta_{11}^2 - \Delta_{41}^2 + \Delta_{11}\Delta}{\Delta + \Delta_{11}}$

$Z_{14} = \dfrac{\Delta_{41}}{\Delta + \Delta_{11}}$

7.8. $Z_{11}(s) = 1$

Chapter 8

8.1. $P_0/P_L = 1+\omega^4$

8.2. The solution is the circuit of Fig. S.1 (a) or (b), where in (a):
$$L_1 = 2/(2\sqrt{(2)}\pi \times 10^4) \text{ H}$$
$$C_2 = 0.4/(2\sqrt{(2)}\pi \times 10^8) \text{ F}$$
$$C_3 = 0.25C_2, \quad L_4 = 6.25L_1$$
and in (b):
$$C_1 = 0.5/(2\sqrt{(2)}\pi \times 10^8) \text{ F}$$
$$L_2 = 2.5/(2\sqrt{(2)}\pi \times 10^4) \text{ H}$$
$$L_3 = 4L_2, \quad C_4 = 0.16C_1$$

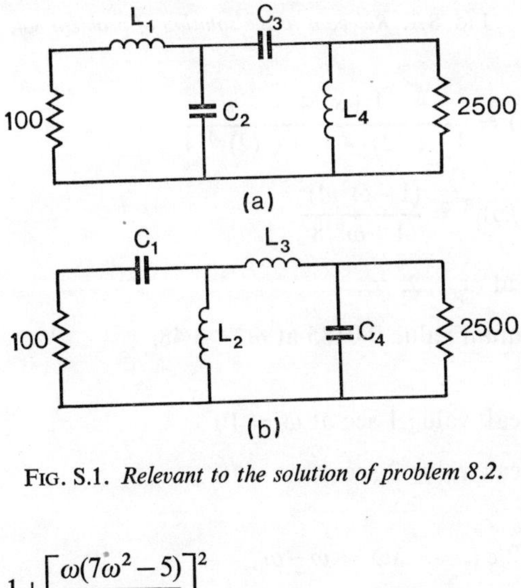

FIG. S.1. *Relevant to the solution of problem 8.2.*

8.3. $\dfrac{P_0}{P_L} = 1 + \left[\dfrac{\omega(7\omega^2 - 5)}{2(5 - \omega^2)}\right]^2$

Insertion loss at $\omega = 10\sqrt{7}$ is $P_0/P_L = 375$, or 25·75 dB.
Insertion loss is 25·75 dB also at $\omega = 2·0$.

8.4. The solution is the circuit of Fig. S.2, with

$$L_1 = \tfrac{1}{2}L_2 = \tfrac{1}{5}L_3 = \frac{50}{4\pi \times 10^7} \text{ H}$$

$$C_1 = 2C_2 = 5C_3 = \frac{1}{\pi \times 10^9} \text{ F}$$

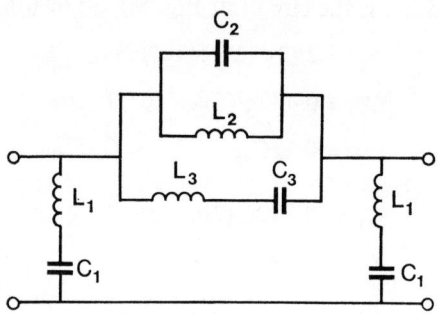

Fig. S.2. *Relevant to the solution of problem 8.4.*

8.5. $Z_{12}(s) = \dfrac{1+s^2/4}{1+\sqrt{(2)}s+s^2+\sqrt{(2)}s^3/4}$

$|Z_{12}(j\omega)|^2 = \dfrac{(1-\omega^2/4)^2}{1+\omega^6/8}$

Zero at $\omega = 2, \infty$.

Minimum value 0·0185 at $\omega = 3\cdot48$.

8.6. (a) Peak value 1 sec at $\omega = 10$

(b) Peak value 2 sec at $\omega = 10$.

8.7. $\tau_c = 2/\sigma_1$, $\Delta\omega = \omega - \omega_0$

$\sigma_1/\omega_0 \ll 1$, and the equaliser must be tuned to the mid-band frequency of the bandpass filter so that $\omega = \omega_0$ corresponds to the centre frequency of the bandpass filter. σ_1 must be chosen to give the correct bandwidth for the equaliser.

8.9. $\tau(\omega) = 10\left(1 + \dfrac{1}{\omega^2}\right) \cdot \tau_0(\omega')$

where $\omega' = 10\left(\omega - \dfrac{1}{\omega}\right)$

and $\tau_0(\omega') = \sqrt{2}\,\dfrac{1+\omega'^2}{1+\omega'^4}$

Approximately, $\tau(\omega) = 20\sqrt{2}\,\dfrac{1+\omega'^2}{1+\omega'^4}$

(It is invalid to approximate further than this.)

8.10. $\tau_c = 2/\sigma_1 = 33\cdot 6$, $\quad \omega_0 = 1$.
In Fig. 8.23(a), $R = 1$, $L = 8\cdot 4$, $C = 1/84$. Values close to these may also give sufficient equalisation.

8.11. For the filter $\tau_f = \dfrac{2(1+\omega^2+\omega^4/4)}{1+\omega^6/8}$

For the equaliser, $\tau_e = 2\cdot 13/(1+1\cdot 133\omega^2)$.

The maximum deviation is $0\cdot 1$ sec.

In Fig. 8.21(a) $R = 1$, $L = 1\cdot 065$.

8.12. Series arms $L_1 = L$ and $C_1 = C$ in parallel, shunt arms $L_2 = R^2 C$ and $C_2 = L/R^2$ in series.
  Therefore $\alpha = 1 - C_1/C_2$, and we must have $C_2 > C_1$ and

$$\frac{C_1}{C_2} + \frac{L_2}{L_1} > 1.$$

Poles and zeros must lie in the two 90° sectors shown as the shaded regions in Fig. S.3.

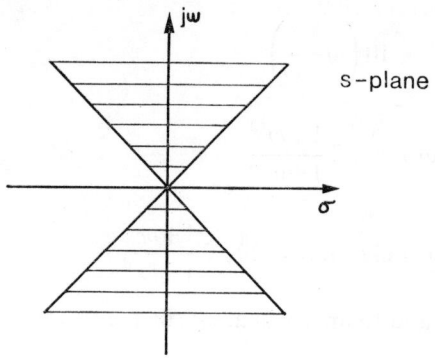

FIG. S.3. *Relevant to the solution of problem 8.12.*

8.13. Same symmetrical lattice as in problem 8.12. Bridged-T realisable without mutual coupling if $R^2C \geqq L$, or $L_2 \geqq L_1$. Bridged-T always realisable if mutual coupling is allowed.

Chapter 9

9.4. (a) $Z(s)+K$ has the same poles as $Z(s)$ and Re $[Z(j\omega)+K] =$ Re $Z(j\omega)+K \geqq 0$ since Re $Z(j\omega) \geqq 0$. Hence the function is p.r.

(b) $Z(s)-K$ has the same poles as $Z(s)$ and Re $[Z(j\omega)-K] =$ Re $Z(j\omega)-K \geqq 0$ if $K \leqq$ minimum of Re $Z(j\omega)$.

(c) $KZ(s)$ has the same poles as $Z(s)$ and its real part is KRe $Z(j\omega) \geqq 0$. Hence the function is p.r.

(d) Re $F(s) > 0$ for Re $s > 0$. Re $Z(s) > 0$ for Re $s > 0$ hence Re $F\{Z(s)\} > 0$ for Re $s > 0$. Thus since $F\{Z(s)\}$ is real for s real it is p.r.

(e) The poles of $Z(s)+s$ are the same as those of $Z(s)$ with a possible additional pole at infinity with residue unity. Also Re $[Z(j\omega)+j\omega]$ = Re $Z(j\omega)$ so the function is p.r.

SOLUTIONS

(f) The poles of $Z(s)-s$ are the same as those of $Z(s)$ if $Z(s)$ has a pole at infinity. In that case the residue at the pole at infinity must be greater than or equal to unity for the function to be p.r. If $Z(s)$ does not have a pole at infinity the function cannot be p.r. owing to the term $-s$.

(g) The poles of the function are given by
$$Z(s) = R(s)+jX(s) = -[(1/s)+s]$$
or
$$R(s) = -\left(\frac{\sigma}{\sigma^2+\omega^2}+\sigma\right)$$
if $s = \sigma+j\omega$.

In the R.H. plane $\sigma > 0$ and $R(s) > 0$ hence there cannot be any R.H. plane poles. On the imaginary axis, $s = j\omega$, and the real part of the function is
$$\frac{R(\omega)}{\{1-\omega^2-\omega X(\omega)\}^2+\omega^2 R^2(\omega)} \geqq 0 \quad \text{since} \quad R(\omega) \geqq 0$$
hence the function is p.r.

9.6. (a) Put
$$Z_1 = \frac{1+\Gamma_1}{1-\Gamma_1}$$
$$Z = \frac{1+\Gamma}{1-\Gamma}$$
then $Z_1 = Z+K$.

Since Γ is bounded real Z is p.r. and from the solution to problem 9.4(a) Z_1 is therefore p.r. Hence Γ_1 is bounded real.

(b), (c), (d)—using similar arguments and the results of problems 9.4(e), (f), (g), the properties of Γ_1 can be discussed.

9.7. (a) No R.H. poles and Re > 0 on $s = j\omega$ hence p.r.

(b) There is a R.H. pole at $s = 1$ hence not p.r.

(c) The degrees of numerator and denominator differ by more than one. Hence not p.r.

(d) The reciprocal of the function is $s + (3/s) + (2/s^2)$. Hence not p.r. owing to double ordered pole at $s = 0$.

9.8. (a) $(S) = \left[1 + \begin{pmatrix} R_1^{-\frac{1}{2}} & 0 \\ 0 & R_2^{-\frac{1}{2}} \end{pmatrix}(z)\begin{pmatrix} R_1^{-\frac{1}{2}} & 0 \\ 0 & R_2^{-\frac{1}{2}} \end{pmatrix}\right]^{-1}$

$\times \left[\begin{pmatrix} R_1^{-\frac{1}{2}} & 0 \\ 0 & R_2^{-\frac{1}{2}} \end{pmatrix}(z)\begin{pmatrix} R_1^{-\frac{1}{2}} & 0 \\ 0 & R_2^{-\frac{1}{2}} \end{pmatrix} - 1\right]$

(b) $(S) = \left[1 + \begin{pmatrix} R_1^{\frac{1}{2}} & 0 \\ 0 & R_2^{\frac{1}{2}} \end{pmatrix}(y)\begin{pmatrix} R_1^{\frac{1}{2}} & 0 \\ 0 & R_2^{\frac{1}{2}} \end{pmatrix}\right]^{-1}$

$\times \left[1 - \begin{pmatrix} R_1^{\frac{1}{2}} & 0 \\ 0 & R_2^{\frac{1}{2}} \end{pmatrix}(y)\begin{pmatrix} R_1^{\frac{1}{2}} & 0 \\ 0 & R_2^{\frac{1}{2}} \end{pmatrix}\right]$

9.9. $S_{11} = S_{22} = \dfrac{-s^3}{s^3 + 2s^2 + 2s + 1}$

$S_{12} = S_{21} = \dfrac{1}{s^3 + 2s^2 + 2s + 1}$

$|S_{12}(j\omega)|^2 = \dfrac{1}{1 + \omega^6}$

9.10. $S_{11} = -S_{12} = \dfrac{n^2 - 1}{n^2 + 1}$

$S_{12} = \dfrac{2n}{n^2 + 1}$

Chapter 10

10.2. Referring to Fig. S.4(a):

(a) $C_1 = \frac{8}{3}$; $L_1 = \frac{3}{8}$; $C_2 = 4$; $L_2 = \frac{1}{12}$; $C_3 = \frac{8}{3}$; $L_3 = \frac{3}{40}$

(b) $C_1 = \frac{15}{8}$; $L_1 = \infty$; $C_2 = 6$; $L_2 = \frac{1}{18}$; $C_3 = \frac{10}{3}$; $L_3 = \frac{3}{50}$

(c) $C_1 = 0$; $L_1 = 1$; $C_2 = \frac{2}{3}$; $L_2 = \frac{3}{2}$; $C_3 = 2$; $L_3 = \frac{1}{6}$

Referring to Fig. S.4(b):

(a) $C_1 = 1$; $L_1 = 0$; $C_2 = \infty$; $L_2 = \frac{8}{15}$; $C_3 = \frac{3}{8}$; $L_3 = \frac{4}{3}$; $C_4 = \frac{3}{32}$; $L_4 = \frac{8}{3}$

(b) $C_1 = 1$; $L_1 = 0$; $C_2 = \frac{3}{4}$; $L_2 = \frac{2}{3}$; $C_3 = \frac{1}{8}$; $L_3 = 2$; $C_4 = L_4 = 0$

(c) $C_1 = \infty$; $L_1 = \frac{8}{3}$; $C_2 = \frac{1}{8}$; $L_2 = 4$; $C_3 = \frac{3}{32}$; $L_3 = \frac{8}{3}$; $C_4 = L_4 = 0$

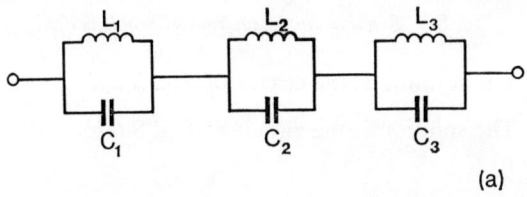

(a)

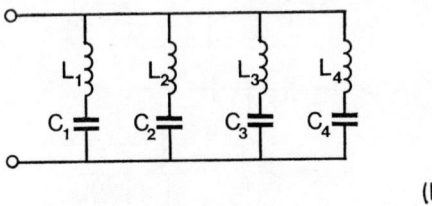

(b)

FIG. S.4. *Relevant to the solution of problem 10.2.*

10.3. By duality it is only required to interchange L and C and series and parallel in the circuits resulting from problem 10.2.

10.4. Referring to the circuit of Fig. S.4(b), $Y_1 - Y_2$ is realised if $C_1 = \infty$; $L_1 = 2$; $C_2 = \frac{1}{3}$; $L_2 = 3$; $C_3 = \frac{1}{24}$; $L_3 = 6$; $C_4 = L_4 = 0$

10.5. $C_1 = 2$; $L_1 = \frac{1}{2}$; $C_2 = \frac{1}{4}$; $L_2 = 1$; $C_3 = 1$; $L_3 = 2$.

10.6. Referring to the circuit of Fig. S.5
(a) $C_1 = 1$; $L_1 = \frac{1}{3}$; $C_2 = 3$; $L_2 = \frac{1}{6}$; $C_3 = 12$; $L_3 = \frac{4}{5}$
(b) $C_1 = 1$; $L_1 = \frac{1}{2}$; $C_2 = \frac{4}{5}$; $L_2 = \frac{25}{6}$; $C_3 = \frac{3}{40}$; $L_3 = \infty$
(c) $C_1 = 0$; $L_1 = 1$; $C_2 = \frac{1}{2}$; $L_2 = \frac{4}{3}$; $C_3 = \frac{3}{2}$; $L_3 = \frac{1}{3}$

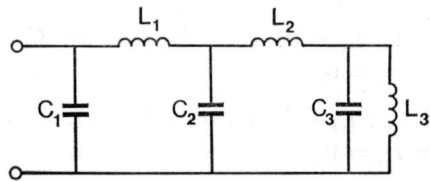

FIG. S.5. *Relevant to the solution of problem 10.6.*

10.7 (a) The solution is the circuit of Fig. S.6(a)
(b) The solution is the circuit of Fig. S.6(b).

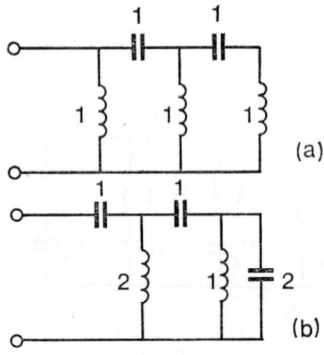

FIG. S.6. *Relevant to the solution of problem 10.7.*

10.8. By duality, it is only necessary to interchange L and C and series and parallel in the circuits resulting from problem 10.6.

10.9. Referring to Fig. S.7(a):

(a) $L_p = 1$; $L_s = 4$; $M = 2$; $C = \frac{1}{2}$

(b) $L_p = 2$; $L_s = 4$; $M = 2$; $C = \frac{1}{2}$

(c) The solution is the circuit of Fig. S.7(b)

(d) Referring to Fig. S.7(a): $L_p = 1$; $L_s = 4$; $M = -2$; $C = \frac{1}{2}$

(e) The solution is the circuit of Fig. S.7(c).

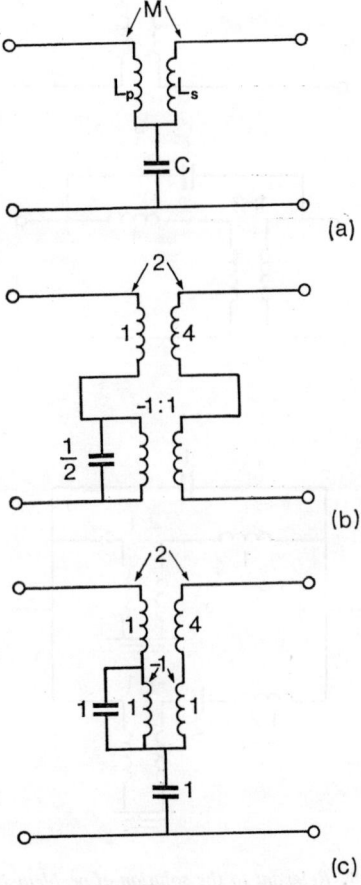

FIG. S.7. *Relevant to the solution of problem 10.9.*

10.10. (*a*) The solution is the circuit of Fig. S.8(*a*)

(*b*) The solution is the circuit of Fig. S.8(*b*)

(*c*) The solution is the circuit of Fig. S.8(*c*).

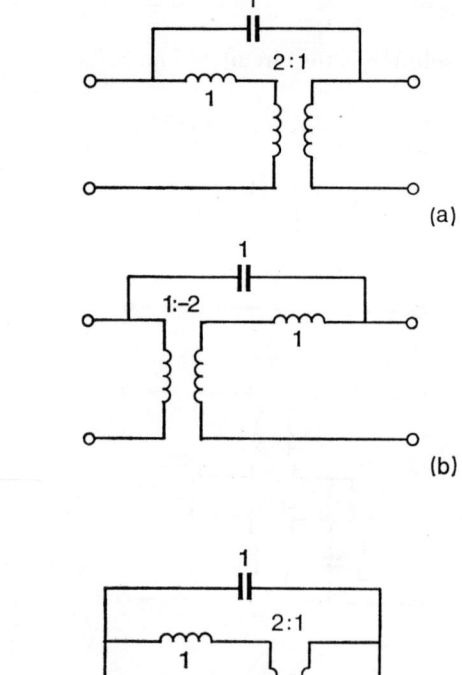

Fig. S.8. *Relevant to the solution of problem 10.10.*

10.11. (i) Referring to the circuit of Fig. S.9(a)

(a) $R_1 = 1$; $C_1 = 0$; $R_2 = \frac{15}{4}$; $C_2 = \frac{4}{15}$; $R_3 = \frac{1}{20}$; $C_3 = 4$

(b) $R_1 = 1$; $C_1 = 0$; $R_2 = \infty$; $C_2 = \frac{3}{8}$; $R_3 = \frac{1}{9}$; $C_3 = 3$

(c) $R_1 = \frac{5}{8}$; $C_1 = \frac{8}{5}$; $R_2 = \frac{1}{40}$; $C_2 = 8$; $R_3 = \frac{1}{28}$; $C_3 = 4$

(d) $R_1 = \infty$; $C_1 = \frac{15}{8}$; $R_2 = \frac{1}{18}$; $C_2 = 6$; $R_3 = \frac{3}{50}$; $C_3 = \frac{10}{3}$

(ii) Referring to the circuit of Fig. S.9(b)

(a) $R_1 = \frac{24}{5}$; $C_1 = \infty$; $R_2 = \frac{8}{3}$; $C_2 = \frac{3}{32}$; $R_3 = \frac{12}{5}$; $C_3 = \frac{5}{72}$; $R_4 = C_4 = 0$

(b) $R_1 = 2$; $C_1 = \frac{1}{4}$; $R_2 = 2$; $C_2 = \frac{1}{8}$; $R_3 = C_3 = R_4 = C_4 = 0$

(c) $R_1 = 0$; $C_1 = 1$; $R_2 = \frac{24}{35}$; $C_2 = \infty$; $R_3 = \frac{8}{9}$; $C_3 = \frac{9}{32}$; $R_4 = \frac{12}{5}$; $C_4 = \frac{5}{72}$

(d) $R_1 = 0$; $C_1 = 1$; $R_2 = \frac{2}{3}$; $C_2 = \frac{3}{4}$; $R_3 = 2$; $C_3 = \frac{1}{8}$; $R_4 = C_4 = 0$.

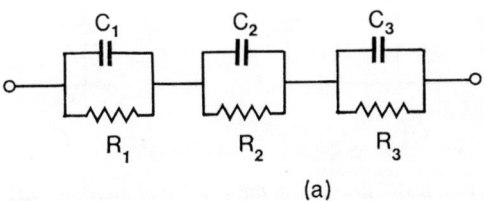

(a)

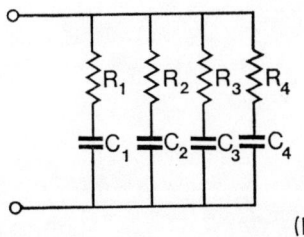

(b)

FIG. S.9. *Relevant to the solution of problem 10.11.*

10.12. (a) The solution is the circuit of Fig. S.10(a)

(b) The solution is the circuit of Fig. S.10(b)

(c) The solution is the circuit of Fig. S.10(c)

(d) The solution is the circuit of Fig. S.10(d).

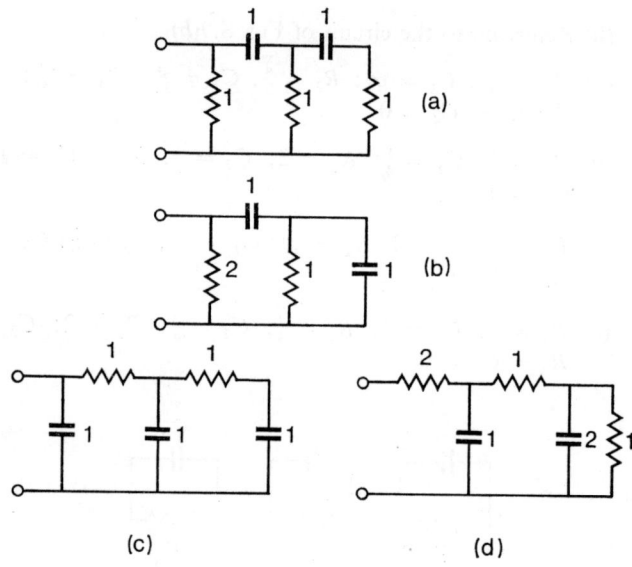

FIG. S.10. *Relevant to the solution of problem 10.12.*

SOLUTIONS

10.13. (a) $Z_{RL}(p) = \dfrac{p(p+2)(p+4)}{(p+1)(p+3)(p+5)}$

(b) $Z_{RL}(p) = \dfrac{(p+2)(p+4)}{(p+3)(p+5)}$

(c) $Z_{RL}(p) = \dfrac{p(p+2)(p+4)}{(p+1)(p+3)}$

(d) $Y_{RL}(p) = \dfrac{(p+2)(p+4)}{(p+1)(p+3)(p+5)}$

(e) $Y_{RL}(p) = \dfrac{(p+2)(p+4)}{p(p+3)(p+5)}$

(f) $Y_{RL}(p) = \dfrac{(p+2)(p+4)}{(p+1)(p+3)}$

In each case the circuit resulting from the corresponding solution to problems 10.2 and 10.3 is modified by replacing each capacitor by a resistor with $R = 1/C$ and leaving the inductors unaltered.

10.14. (a) $Z_{RL}(p) = \dfrac{p(p+4)(p+6)}{(p+1)(p+5)}$

(b) $Z_{RL}(p) = \dfrac{(p+2)(p+4)}{p+3}$

(c) $Z_{RL}(p) = \dfrac{p(p+4)(p+6)}{(p+1)(p+5)(p+7)}$

(d) $Z_{RL}(p) = \dfrac{(p+2)(p+4)}{(p+3)(p+5)}$

In each case the corresponding circuit resulting from the solution to problem 10.11 is modified by replacing each resistor by an inductor with $L = R$ and replacing each capacitor by a resistor with $R = 1/C$.

10.15. (a) The solution is the circuit of Fig. S.11(a)

(b) The solution is the circuit of Fig. S.11(b)

(c) The solution is the circuit of Fig. S.11(c)

(d) The solution is the circuit of Fig. S.11(d)

(e) The solution is the circuit of Fig. S.11(e).

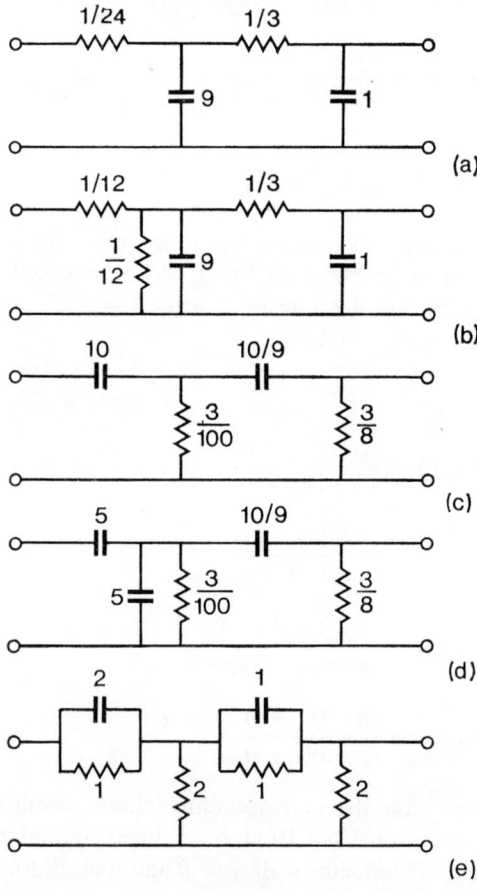

FIG. S.11. *Relevant to the solution of problem 10.15.*

10.16. The solution is the parallel connection of the three circuits in Fig. S.12 and the value of K is 521/24.

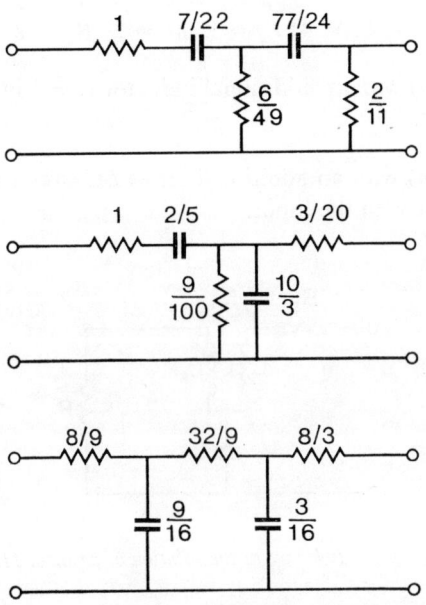

FIG. S.12. *Relevant to the solution of problem 10.16.*

Chapter 11

11.1. $0 \leq \gamma \leq 4$

The required circuit values are: $C = 1$; $R = 1/(4-\gamma)$; $r = 1/(\gamma^2 - 4\gamma + 8)$; $L = \gamma/(\gamma^2 - 4\gamma + 8)$.

11.2. Referring to the circuit of Fig. S.13:

(a) $R_1 = 2; L_p = 2; L_s = 8; M = 4; R_2 = 1$

(b) $R_1 = 1; L_p = 4; L_s = 1; M = 2; R_2 = 2$

(c) As (a) with an additional inductor $L = 1$ in series with the input.

(d) As (a) with an additional series LC circuit $L = C = 1$, in shunt with the input.

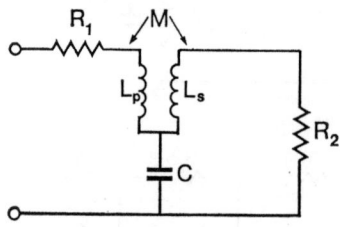

FIG. S.13. *Relevant to the solution of problem 11.2.*

11.3. Referring to the circuit of Fig. S.14(a):

(a) $L_p = 4; L_s = 1; M = 2; C = \frac{1}{4}$

(b) $L_p = 1; L_s = 1; M = -1; C = 1$

(c) $L_p = \frac{2}{3}; L_s = \frac{8}{3}; M = -4/3; C = 12$

(d) The solution is the circuit of Fig. S.14(b)

(e) As (b) with the addition of an inductor, $L = 2$, in series with the input (*f*) The solution is the circuit of Fig. S.14(*c*)

(*g*) The solution is the circuit of Fig. S.14(*d*)

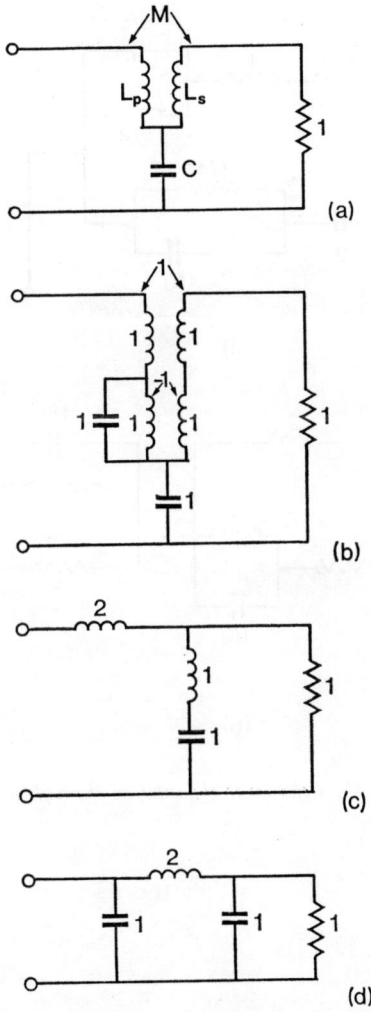

Fig. S.14. *Relevant to the solution of problem 11.3.*

11.4. (*a*) The solution is the circuit of Fig. S.15(*a*)

(*b*) The solution is the circuit of Fig. S.15(*b*).

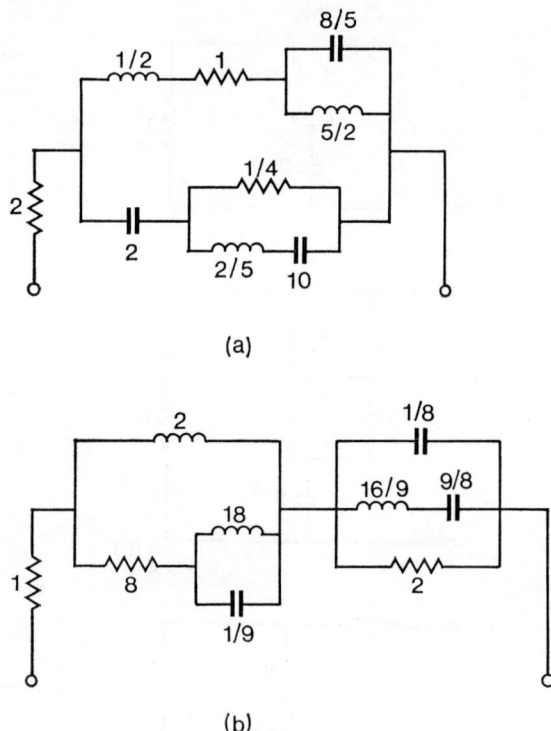

FIG. S.15. *Relevant to the solution of problem 11.4.*

Chapter 12

12.1. The solutions to parts (*a*), (*b*), (*c*), (*d*) are the circuits of Fig. S.16(*a*), (*b*), (*c*), (*d*) respectively.

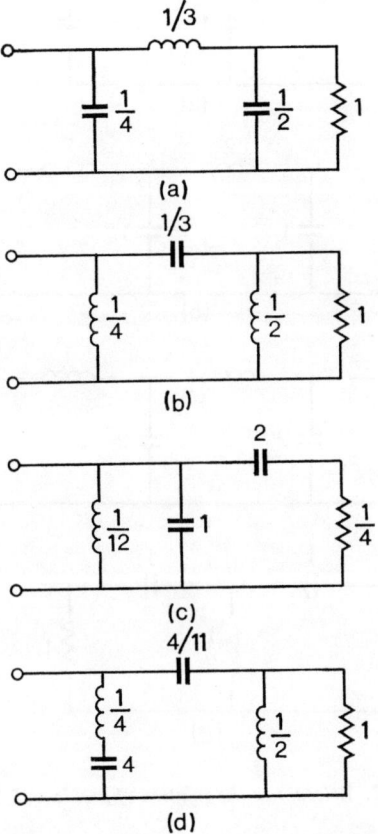

FIG. S.16. *Relevant to the solution of problem 12.1.*

12.2. The solutions to parts (*a*), (*b*), (*c*), (*d*) are the circuits of Figs. S.17(*a*), (*b*), (*c*), (*d*) respectively.

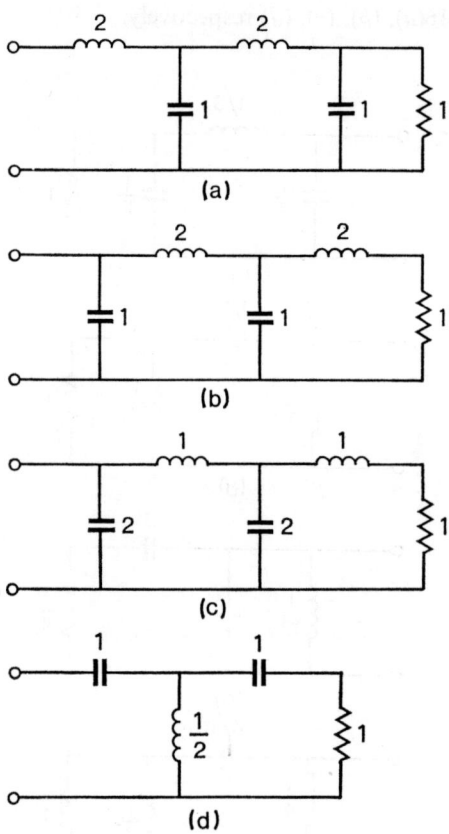

FIG. S.17. *Relevant to the solution of problem 12.2.*

12.3. The solutions to parts (*a*), (*b*), (*c*), are the circuits of Figs. S.18 (*a*), (*b*), (*c*), respectively or the corresponding dual configurations.

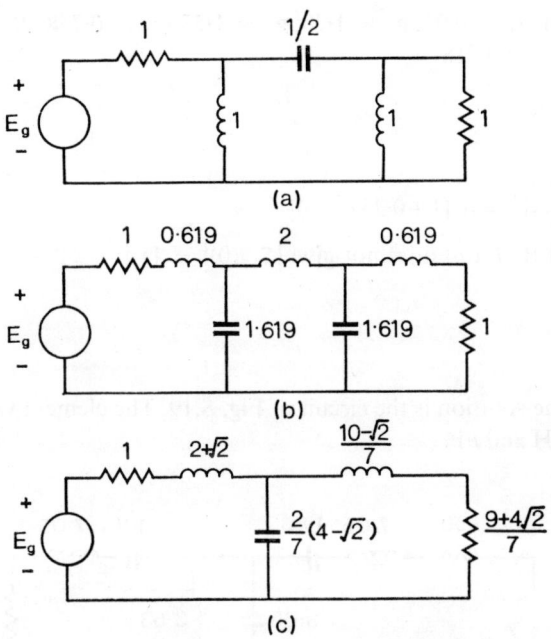

Fig. S.18. *Relevant to the solution of problem 12.3.*

12.4. (*a*) $|S_{12}|^2 = 1/(1+\omega^8)$

(*b*) $|S_{12}|^2 = \dfrac{1}{1+\{\omega/[(6\pi)(10^3)]\}^8}$

(*c*) $|S_{12}|^2 = \dfrac{8/9}{1+\omega^8}$

(*d*) $|S_{12}|^2 = \dfrac{1}{1+(\omega/1\cdot 12)^{12}}$

12.5. The solutions are ladder networks of the type shown in Fig. 12.18 with

(a) $g_1 = 0.85$, $g_2 = 1.1$, $g_3 = 0.85$, $R_L = 1$

(b) $g_1 = 0.93$, $g_2 = 1.28$, $g_3 = 1.57$, $g_4 = 0.758$, $R_L = 1.221$ or 0.818

12.6. $|S_{12}|^2 = 1/[1 + 0.25 T_5^2(\omega)]$

(N.B. $T_4(\omega)$ does not give $|S_{12}(0)| = 1$)

12.7. The solution is the circuit of Fig. S.19. The element values are mH and µF.

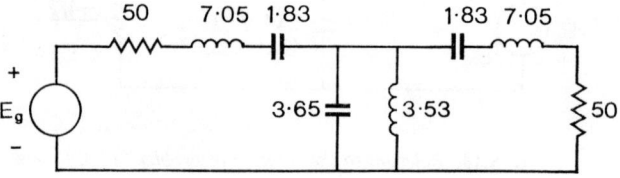

FIG. S.19. *Relevant to the solution of problem 12.7.*

12.8. (a) 1·76 (b) 0·065 secs (c) 4·37 (d) maximally flat insertion loss $|S_{12}|^2 = 1/65$; maximally flat delay $|S_{12}|^2 = 1/16$.

12.9. $\left(1 - \dfrac{\omega_0^2}{p^2}\right) T_g \left(p + \dfrac{\omega_0^2}{p}\right)$

12.10. (a) The solution is the circuit of Fig. S.20(a)

(b) The solution is the circuit resulting from problem 10.9(e) terminated at port 2 in a 1 ohm resistor

(c) The solution is the circuit of Fig. S.20(b).

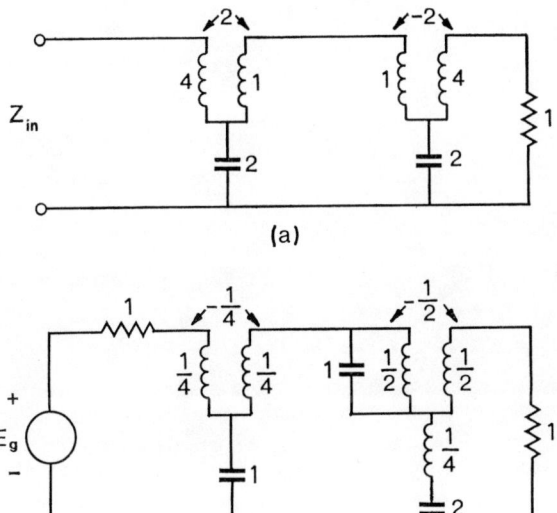

FIG. S.20. *Relevant to the solution of problem 12.10.*

12.11. (a) $RC\omega_c = 2$

(b) $RC\omega_c = 9\cdot08$

(c) $RC\omega_c = 5\cdot5$

(d) $RC\omega_c = 6\cdot9$

12.12. (a) $RC\omega_c = \left(\dfrac{3R^2C}{L} - 1\right)^{\frac{1}{2}}$

$$\ln(1-T)^{-\frac{1}{2}} = \pi \bigg/ \left(\dfrac{3R^2C}{L} - 1\right)^{\frac{1}{2}}$$

12.10 (a) The solution is the circuit of Fig. S.200.

(b) The answer is the circuit resulting from problem 11.9(a) terminated at port 2 into 1 ohm resistor.

c. The solution is the circuit of Fig. S.20lb.

Fig. S20. Realizable Networks for problem 12.10.

12.11. (a) $R = ?$

(b) $RC = 9.0s$

(c) $RC = 4.5$

(d) RC

12.12. for $R_0 C_0 = \left(\frac{1}{T-1}\right)$

$R_0(T-1) = \left(\frac{2R_1 C_1}{T_2}\right)$

INDEX

Allpass networks, 354, 359, 520
   C-Section, 361
   D-Section, 363
Approximation problem, 492
Augmentation, 464

Balanced bridge, 471
Bandpass approximations, 511
   general form, 512
   lowpass prototype, 513
Bandpass filter, 326, 329
   negative frequencies, 331
Bandstop filter, 333
Bandwidth conservation, 332
Bessel polynomial, 518
Bilateral elements, 304
   networks, 371
Bott and duffin synthesis, 468
Bounded real function, 383
Broadband matching theory, 531
   of RLC load, 541
   of RC load, 534
   with maximally flat response, 536
   with chebyshev response, 537
Brune preamble, 450, 461, 469
   section, 423, 456, 525
   synthesis procedure, 449
   extraction cycle, 456
Butterworth approximation (*see* Maximally flat approximation)

C-Section, 423, 445, 525
Cascade synthesis, 521
   allpass networks, 530
   explicit formulae, 529
   sections, 523
   sub-networks, 522
Canonical circuit, 400
Cauer forms, 404

Chebyshev approximation, 501
   detail, 505
   element values, 511
   general behaviour, 502
   recurrence formulae, 507
   roots, 508, 510
Compact network, 417, 420
Complex conjugate roots, 310
Complex frequency, 285
Complex frequency plane, 285
   imaginary frequencies, 287
   real frequencies, 287
Continued fraction expansion, 410, 483
   generation of coefficients, 413
Cut-off frequency, 326

D-Section, 445, 446, 525
Darlington procedure, 487
   synthesis, 461
   theorem, 461
Decrement (*see* Fractional bandwidth)
Delay approximation, 513
   maximally flat, 516
   recurrence formula, 517
Delta function, 311, 313
Denormalisation, 323
Determinant (evaluation of), 297
   Cramer's rule, 298
Distortion, 348
Double-terminated network, 477, 486
Driving point functions, 306, 309, 311
   impedances, 376

Elliptic function response, 519
   realisation, 520
Energy functions (of a network), 379
Energy in RLC networks, 372
   dissipated, 374, 375
   stored, 374, 375

INDEX

Equiripple approximation (*see* Chebyshev approximation)
Equivalent circuits, 338
Extraction (from impedances), 403

Foster forms, 400
Foster function, 384
Foster section, 334
Four-port network, 321
Fourier transform, 352
Fractional bandwidth, 332
Frequency scaling, 323
Frequency transformations, 326

Geometrical symmetry, 331
Group delay, 348, 350
   equalisation, 362, 367
   of bandpass filter, 369
   of a two-port, 353
Group velocity, 350, 352
Guillemin's synthesis procedure, 441, 448

High pass filter, 328
Hurwitz polynomial, 384, 462

Ideal transformer, 323, 334, 336
Impedance matrix, 418
Impedance scaling, 291, 323
Impedance transformations, 334
Impedance transforming filters, 343
Initial conditions, 286
Insertion loss, 326, 330, 349
Insertion phase (*see* Transfer phase)

LC transfer functions, 415
Ladder networks, 488
   continued fraction expansion, 489
Laplace transform, 285
Lattice networks, 355
   constant resistance, 358
   image parameters, 359
Linear networks, 371, 372
Loop analysis, 296
Loop impedance matrix, 301
Loop resistance matrix, 374

Maximally flat approximations, 485, 493
   element values, 501
   roots, 495
Mesh analysis (*see* Loop analysis)
Minimal realisation, 400
Minimum reactance function, 451, 462
Minimum susceptance function, 451, 462
Mutual admittance, 305
Mutual impedance, 300

N-Loop circuit, 301
Natural frequency, 288
Necessary conditions, 371
Nodal analysis, 303
Normalisation, 323
Null network, 380

Open-circuit driving point impedance, 306

Pi-network, 342
Partial fractions, 312
   expansion, 394
Partial pole removal, 440
Passband, 492
Passive networks, 371
Phase equalisation, 355
Phase velocity, 351, 352
Pole of attenuation, 345
Poles (definition), 289
Poles and zeros of RL circuit, 291
   of RLC circuit, 293
Pole-zero plots, 291
Port normalising number, 387
Positive real function (definition), 380
   Imaginary axis poles, 381
   Poles and zeros, 380
   Reciprocal, 380
Positive semi-definite matrix, 374, 375, 378
Private poles, 416
Prototype networks, 326, 492

Q-factor, 332
Quadratic form, 374

INDEX

RC driving point impedance, 423
 partial fraction expansion, 426
RC transfer functions-transmission zeros, 435
RC-LC transformation, 431
RL driving point impedance, 433
RLC driving point impedances, 448
 partial fraction expansion, 448
Radar, 348
Rational functions, 309
Reactance function, 384, 393
 poles and zeros, 385
 slope, 396
Resonant frequency of RLC circuit, 294
Richards' theorem, 468

Scaling-effect on time domain response, 324
Scaling of networks, 323
Scattering parameters, 372
 of a two-port, 386
Self admittance, 305
Self impedance, 300
Short-circuit driving point admittance, 307
Short-circuit transfer admittance, 307
Single-terminated network, 472, 481
Stopband, 492

Sufficient conditions, 371
Synthesis, 371

T-network, 343, 345
Television, 348
Transducer power gain, 387, 480
 relation to scattering matrix, 388
Transfer function, 306, 309, 313
 synthesis, 477
Transfer impedance, 307
Transfer matrix, 348
Transfer phase, 348
Transformerless networks, 335
Transition region, 492
Three-port network, 320
Tuned circuits, 331
Twin-T network, 370
Two-element-kind networks, 393

Voltage transfer function, 349
Voltage transfer ratio, 434

Wheatstone bridge, 303

Zeros (definition), 289
Zeros of transmission, 479